THE COATTAILS OF GOD

Also by Robert M. Powers

Planetary Encounters
Shuttle: The World's First Space Ship

THE COATTAILS OF GOD

The Ultimate Spaceflight— The Trip to the Stars

ROBERT M. POWERS

A Warner Communications Company

Warner Books, Inc., 75 Rockefeller Plaza, New York, N. Y. 10019
A Warner Communications Company
Distributed in the United States by Random House, Inc., and
in Canada by Random House of Canada, Ltd.
Printed in the United States of America
First Printing: November 1981
10 9 8 7 6 5 4 3 2 1

Book design by Judy Allan (The Designing Woman)
Text illustrations by Ron Miller

Library of Congress Cataloging in Publication Data
Powers, Robert M., 1942–
The coattails of God.

Bibliography: p.
Index.
1. Interstellar travel. I. Title.
TL793.P68 919.9'04 81-3001
ISBN 0-446-51231-1 AACR2

For Michelle Powers, daughter, with love.

"Have you ever seen a rocket taking off? It stands there, so big that the men around it look tiny, like flies; they light a little spark, and a great bellow tears the air to shreds, a white cloud spreads, and it lifts off, up into the infinite, and you blaspheme: God, we've caught you by the coattails!"

RAY BRADBURY, to Oriana Fallaci, 1966

ACKNOWLEDGMENTS

I would like to thank my editor, Fredda Isaacson, who was a constant source of encouragement throughout a longer time than either of us imagined. My thanks to Betty Chancellor, who helped with editing some of the early stages of the manuscript, and to Michele Hughes, who spent long hours assisting in some of the research.

The British Interplanetary Society was kind enough to allow material from their Project Daedalus to be described, and NASA provided information used on current and near-future technologies for space. I am indebted to the participants of the 1977 Summer Study at Ames Research Center for their thorough work on space resources and space settlements, which appeared in 1979 as a NASA technical paper. I owe an equal debt to the participants of the 1975 Summer Study on space settlements sponsored by Stanford University, the American Society for Engineering Education, Ames Research Center, and NASA; this material appeared in book form from NASA's scientific and technical office in 1977.

Other institutions provided access to historical and older technical information, including the Engineering Library of the University of Colorado. Fearon-Pitman Publishers, Inc., provided a copy of their comprehensive work *Rocket Propulsion and Spaceflight Dynamics,* by J. W. Cornelisse, H. F. R. Schöyer, and K. F. Wakker, which was useful.

A number of people looked at the manuscript in various stages and commented in interesting and helpful ways, including John Harrison, Michele Cser, and others. Dr. Thomas Heppenheimer read the entire book in manuscript and commented at length on some portions. Isaac Asimov wrote to clarify a few details about tachyons, and to make some remarks about the General Unification Theory.

I wish to thank numerous scientists and others who have speculated, dreamed, and written about starflight over the course of this century. Most of their work can be located in the Reference section. It is a tribute that they have worked in technical detail on paper with a subject that to many of them at the time must have seemed centuries in the future.

I hope that in reading this discussion about going to the stars, you, the reader, see it as the great dream it is. If enough people see it so, it will be an inevitable dream.

ROBERT M. POWERS
Los Angeles

CONTENTS

PREFACE

The Great Ring of Brogar lies in the Orkney Islands, north of Scotland. It stands in a commanding position on the peninsula that separates Loch of Harray from Loch of Stenness on the mainland of Orkney. The giant stones of Brogar, flat slabs fifteen feet high, can be seen from great distances at sea. To the west of the Ring, the ground rises steeply. The circle of stones and the small cairns close by were very precisely placed on the island; the Ring is within two inches of a perfect 125 megalithic yards (about 2.720 feet per yard) in diameter. The people who built it must have spent years moving the huge stones to correspond exactly with their ancient measurement. The whole arrangement lies near the 59th parallel; this ancient people was able to locate the overhead meridian with an error of less than nine minutes of arc, a third the apparent diameter of the moon, in the second millennium B.C.

There is no convenient explanation why ancient man volunteered to face the harsh winters on the northern outposts of Scotland to build an edifice which must have taken generations to construct and maintain. We know now they wished to observe the moon. Was that all? We know a little more about the great observatory at Stonehenge, whose several stages of building took a thousand years, and some details of the installation at Er Grah near Carnac. We have dug and picked at Temple Wood and Mid Clyth. There are dozens of other strange and near-vanished sites of similar construction in Ireland.

The building of the Pyramids, a better-known enterprise of antiquity, and the giant stone statues of Easter Island both took the work of generations to complete. Their builders must have been sustained in some way by the dream of what they were constructing. But the feelings of those times, the timber and lath of their dreams, are gone now. All that remains are the decaying apparitions scattered about the corners of the world, our present curiosity about them, and the thought that our ancestors perceived a great purpose, at different times and in different parts of the world.

Man's efforts at cathedral-building, the sustained thrust that brought about democracy in a large part of the world, the constant expansion of human health care, and the sustaining of an industrial revolution and technology are similar events which required the attention of several generations. That mankind should again undertake a gigantic project, the goal of which extends beyond the lifetimes of the originators, would therefore not be historically surprising. Voyages to the stars are projects of that magnitude. We will build a starship someday not because we are preordained to wander in the cosmic nothingness, but because we are planners and practical explorers and capable of sustained drives covering more than our lifetimes. We are dreamers, too, and the stars are a wonderful dream.

In the spring of 1917, Robert H. Goddard handed over a thick envelope full of papers to a friend for safekeeping. "This stuff," he said, "is to be read only by an optimist." Among the papers discussing solar energy, ideas about atomic power, and speculations on how to avoid meteor collisions in outer space, there was one entitled "The Ultimate Migration." "It has to do," Goddard told his friend, "with what will surely be the last pioneering effort of the human race. I truly believe that it will eventually be possible for human beings to be transported on expeditions to the nearest stars—leaving earth while they are in a kind of deep sleep, the way seeds 'sleep' over the winter, and waking up at their destinations as much as a thousand years later!"

Goddard went on to build the first liquid-fuel rocket, and it was by using his hundreds of patents on rocket technology that mankind managed, decades after his death, to reach the moon—a goal first dreamed of by the Greeks. Robert Goddard was one of those whose vision and knowledge advance the boundaries of our horizons: the planners, builders, and dreamers.

In 1973, more dreamers and planners constructed two unmanned spacecraft to study Jupiter and Saturn. Though these vehicles were not originally designed to do so, they will become the first starprobes. NASA's *Pioneer 10* and *Pioneer 11* will both eventually exit the solar system and fly toward the stars—very slowly. Even though both of these spacecraft are among the fastest objects mankind has ever devised, it will be 250,000 years before *Pioneer 10* reaches the vicinity of another solar system circling another star. It is now headed for a point on the boundary between the constellations Taurus and Orion.*

Had either spacecraft been aimed at the *nearest* star when it was launched, it would still take more than 100,000 years to reach its destination. If this were the best mankind could do, starflight would be impossible.

But the notion of a genuine starship is an old one. Like the Martians in Nigel Kneale's *Quatermass*, "starship" is a name "almost worn out before anything arose to claim it." The earliest rocket pioneers thought about starships and they were followed by scores of other dreamers who put their strange designs on paper. Though no starship has ever been built, discounting for the moment *Pioneer 10* and *11*, there are plenty of people who think one can eventually be built and probably will be. Some think it will be only a matter of time, perhaps only a century, perhaps less, given the right scientific and technological breakthroughs. Others regard it as a much more distant and complex problem.

Starflight is distinctly unlike other technologies which have developed in a few short decades—automobiles, aircraft, space rockets, computers—and different from some otherwise historically similar engagements of the human mind. Most comparisons of the events of aviation or automotive history with the progress of space technology miss an important point. Autos and aircraft, electronics and microcomputers—even DNA research—were developed in entrepreneurial circumstances. At a relatively early stage of their development, the products were practical for widespread use and could be delivered to mass markets. This, of course, stimulated their further technical advance. By contrast, application of space and nuclear technologies has been costly, involving tremendously complex development, and has been only

*On July 26, 1981, Pioneer 10 was 25 Au distant from the sun—2,323,895,000 miles (3,739,947,300 kin).

partially and slowly applied to mass markets. These technologies are principally investigated by gigantic organizations and have remained, for the most part, the sole property of bureaucracies the size of governments.

Also the sequence of events which starts with Columbus and continues through the *Mayflower*, the *Great Eastern*, the *Queen Elizabeth*, the Stratocruiser, the 707, and the Concorde represents an *increasing* cost series, not one which decreases. While it is true that the Viking mission to land an unmanned spacecraft on the surface of Mars cost less per capita than the average consumption of bubble gum in the United States for a few years, and the moon landing could have been financed by the sacrifice of a case of beer for each American for five years or so; the cost of starflight, if it happens, will be utterly staggering. It will be as unlike the moon landing in economics as the annual budget of a dying hamlet is unlike the *world* GNP.

But starflight will come someday. Technology has a way, comparisons notwithstanding, of bringing fantastic changes in a relatively short time. Who would have thought the doodlings of old Leonardo would have evolved from ideas on vellum to the reality of Concorde in a few short centuries? Had our grandfathers been faced with the prospect of sending a vehicle out to a point two billion miles distant, they would have been defeated; they had no technology for such a voyage. But there are people alive today who have seen both the flight of one of the first aircraft and the photographs of Saturn (on the six-o'clock TV news), a quantum jump in speed from 60 mph to over 25,000 mph.

At this time, however, the space program is sliding into a decline. Doom is predicted from every corner. We reach with a collective longing for a world in which our cars are made safe beyond our ability to drive them; we want to be protected from any remote chance of unforseen effects in our medicine, our food, the environment.

Some do not want the risk of nuclear energy. The underlying idea of "no risk" is in itself dangerous. No progress was ever made without risk, both individual and collective.

We have lost the gods of our tribe; our wise men no longer set forth noble goals. They speak instead of darkness and turmoil. It is time we remembered the stars again. Though we in this generation will never reach them, someday one of our number will.

Even if the starship we envision becomes as laughable to some future man as Jules Verne's cannon shot to the moon is to us, that is no reason to abandon dreaming or planning or designing. We did not get to the moon by being afraid of the magnitude of the project or of the difficulties involved.

That the Vikings could not cross the ocean in a few hours (as they would have been able to do a thousand years later) did not stop them from crossing nonetheless. That it may take us two generations, or three, or fifty, to reach the stars should not stop us, either. Someone has to try out the idea, regardless of how primitive the effort may be. Otherwise, dreams remain only dreams.

"Wonders are many, and none is more wonderful than man; the power that crosses the white sea, driven by the stormy south-wind, making a path under surges that threaten to engulf him; and Earth, the eldest of the gods, the immortal, unwearied, doth he wear. . . . Yea, he hath resource for all; without resource he meets nothing that must come; only against death shall he call for aid in vain; but from baffling maladies he hath devised escapes."

SOPHOCLES, *Antigone*

CHINAMEN, CARS, AND MAD YELLOW TEAKETTLES

*"C'est brusque et brutal, mais ça marche!"**

EMILE LEVASSOR, 1890

From the domestication of the horse in antiquity until the middle of the nineteenth century, speed was limited by the capability of four legs and muscle. Muscle power was the only motive force available; and as a propulsion system it was primitive, inefficient, and produced, not surprisingly, only one horsepower.

*"It's choppy and rough, but it works!"

Though there have been thousands of arguments about racehorses over the centuries, it is unlikely that a man ever traveled more than 50 mph on one. (Ctesias's remark, in 398 B.C., that unicorns were much faster than horses notwithstanding.) One of the fastest horses, *Sonido,* was clocked at only 40.54 mph over a one-half-mile course on June 28, 1970. In 1973, *Truckle Feature,* a quarter horse owned by Gordon Howell, ran a quarter mile in 21.02 seconds, faster than *Sonido* but still under 50 mph.

This historical "one-horsepower barrier" was breached once, though, by an obscure Mandarin named Wan-Hoo, who about A.D. 1500 designed a rocket airplane, of sorts. It consisted of two large kites connected by a framework, in the center of which he had placed a saddle. Underneath the kites were forty-seven large powder rockets and forty-seven Chinese assistants holding flaming ignition torches. Wan-Hoo climbed aboard his unnamed rocket plane and gave a prearranged signal. He is reported to have disappeared in a cloud of dense black smoke. It is entirely possible that he reached a speed of more than 50 mph at some point during his spectacular suicide.

While the story of Wan-Hoo is of somewhat dubious authenticity, the record of the rise of technology and machinery is not. As soon as a motive force greater than muscle power could be successfully harnessed to a vehicle, the quest for greater speeds began.

The first primitive effort to improve on the horse was undertaken in 1769 by Nicolas Cugnot, a Frenchman. He had no way of knowing that he was a primary domino in a long stretch of invention and technology that would end with a man standing on the moon and other men bending over drafting boards plotting the conquest of the stars. Cugnot had been commissioned to build a vehicle to tow heavy field artillery, and he designed a "steam tractor" to do so. It had the dubious distinction of being in the first auto accident when it ran into a wall on a test run. Two people could ride on it, but it wasn't faster than the animal it was intended to replace.

Steam engines were perfected after Cugnot's tractor, and by 1802 a high-wheeled steam coach survived the trip from Cornwall to London. The steam coach was not greeted with any particular enthusiasm by the public, and owners of even modestly fast horses were not invited to challenge the strange contraption to races.

The steam coach gave way to the steam locomotive, and the position of the horse as the fastest form of transportation crumbled. The *Rocket,* one of the first true steam locomotives, managed 25 mph with one carriage and a few passengers during trials in 1829. By 1840, the *Patentee,* known as a 2-2-2 locomotive, was running on British tracks at nearly 60 mph. By the middle of the nineteenth century a man-carrying vehicle traveled more than a mile each minute.

The train was principally designed for hauling freight, not for setting speed records. There was a limit to how fast a steam-powered train could go. A smaller and lighter power source and a smaller and lighter vehicle were needed.

In 1880, small steam automobiles made their appearance, especially in France, built by Le Compte de Dion, whose name still survives in the design called the "de Dion" rear-axle arrangement. None of these cars could beat a steam locomotive *or* a fast horse. But, thanks to Gottleib Daimler at the Gasmotoren-Fabrik Deutz in Germany, a new power plant was developed, one that would set new speed records in cars and airplanes and would prevail until the invention of the jet engine and the perfection of rockets.

Daimler perfected the first gasoline engines, as did another German, Karl Benz. Both of these automotive pioneers' names remain in the Mercedes-Benz and Daimler cars, the latter built by Jaguar, which bought out the firm in the 1950s. Benz's first gasoline engine produced exactly one horsepower from a single cylinder. Daimler's engines powered a motorcycle, a motorboat, and an early airplane. In 1891, a four-seater Panhard-Levassor automobile claimed a top speed of 18½ mph; within four years there were Daimler engines in Panhard delivery trucks on the streets of Paris.

The first *world* land-speed-record attempt by automobile was made on December 18, 1898: 39.24 mph in an electric car. It was promptly broken at 43.69 mph, also by an electric.

At about this time serious scientific discussion emerged concerning what speeds the human body and engineering contraptions, such as automobiles, could stand. From time to time, a noted scientist or medical expert claimed that the human body could not withstand a certain speed; then someone else with more guts than brains would prove him wrong. It was a controvery which would last into the late 1960s, when it was still occasionally asserted that a human body could not withstand the "g" forces necessary to achieve earth orbit, or could not survive zero-g for extended periods. Some turn-of-the-century scientists were convinced that a man could not breathe at speeds above 70 mph and that an auto going more than 60 mph would rattle itself to pieces. Even Karl Benz thought this latter possibility likely.

The year before the Wright brothers soared over the dunes at Kitty Hawk, a steam car set a world record of 75.06 mph. Predictions were revised. Now they forecast that if the 100-mph barrier were crossed, great harm would come to life, limb, and machine. In 1904, Barney Oldfield nudged the magic 100 barrier in a bare, crude race car called Old 999. He sped across a frozen lake surface near Detroit without a clutch or body on the car, and no brakes, to reach 91.37 mph. Shortly afterward, a Darracq motorcar managed 104.52 mph. In 1909, Oldfield drove a modified Benz Grand Prix car called a

Blitzen Benz to a record 131.7 mph. By the time this speed had been significantly increased by the automobile, a new device called the airplane had far outrun it.

The first aircraft, like the first cars, could not outdo the previous type of machine in speed. But long before the start of World War I, aircraft were traveling at more than 100 mph. By 1923, while the speed record for automobiles was still a year away from being set at 141 mph, an American pilot named Williams flew at 266 mph. Pilots raised their goals. In 1933, an Italian aviator flew his propeller-driven plane at 423 mph.

Mankind was approaching the first *real* barrier since the one-horsepower limit: the speed of sound. And as the sound barrier loomed ahead, the power plant which would eventually break it was just being perfected. Robert Goddard had been experimenting with rockets and rocket engines all through the 1920s. So had several groups in Germany.

On June 11, 1928, the spirit of Wan-Hoo returned. On that day, the first modern man to fly by rocket power took to the air. His name was Friedrich Stamer, and he flew a rocket-powered glider, hardly less primitive than Wan-Hoo's device. (It looked as if it were flying backward.) He flew from the Wasserkuppe, one of the Rhön Mountains in western Germany, for almost a mile, at a speed of 60 mph. Less than ten years later, the Germans were trying out rocket engines in airplanes; and by 1944 a combat aircraft called the Me-163B Komet had been built.

Although the jet airplane had also been invented, the engineers already knew that the rocket plane would be much faster. Unmanned V-2 rockets of the Second World War had pushed the speed record to several thousand miles per hour. In the United States, engineers of the Bell Corporation were asked to design a peacetime research rocket plane in December, 1944. It was designed to test out the parameters, the "envelope," of an aircraft flying faster than sound.

There was increasing talk of a "sound barrier" heard more often in the media than around the campfires of aeronautical engineers. While extension of low-speed aerodynamic theories did *predict* infinite drag at the sonic speed ("infinite drag" meaning the plane would cease to fly faster), the success of the V-2 and other rockets clearly demonstrated that it must not be so. The aeronautical engineers, at least, if not the test pilots and the media, were convinced that there was no "sonic barrier." They were busy building the Bell Corporation's airplane.

Pilots during the war had reported strange experiences as their propeller aircraft approached higher and higher speeds. Some said the aircraft did not "feel right"; others claimed the control surfaces worked in a manner *opposite* to the normal way. Even the famous Dr. Theodor von Kármán, who later founded Jet Propulsion Laboratories, jokingly referred to a "brick wall in the sky."

The first scientist to investigate the speed of sound was Sir Isaac Newton, and a correct theory for sound speed was presented by Laplace early in the nineteenth century. The name of Dr. Ernst Mach, an Austrian physicist who investigated phenomena associated with the speed of sound, was adopted to indicate the ratio of flow velocity to sound speed: the Mach number. The speed of sound was Mach 1, and it varied in relation to altitude, wind speed, and, most particularly, temperature. At sea level on a calm day at average temperature, the speed is about 760 mph. At 40,000 feet, it is 660 mph.

One of the early pioneers in trying to achieve the satisfaction of speeding away from his own laughter was Geoffrey de Havilland, the son of the famous British aircraft engineer and designer. In a DH-108S, de Havilland died trying to reach Mach 1. It was reported that his plane experienced severe buffeting and then disintegrated.

The Bell Corporation's research airplane, called the Bell X-1, was nothing less than a rocket ship. The stubby wings of almost solid metal, for strength, measured only 3½ inches at the thickest part. The engine burned alcohol and liquid oxygen, and firmly bolted to its tail were four rocket tubes.

The engine of the Bell X-1 had been the subject of some worry. It was much smaller than most people who saw it had expected it to be. That it could develop more than 24,000 horsepower from a 210-pound weight was doubted by quite a few. A guard at the Bell plant was quite succinct: "Everybody's all het up about a toy," he muttered. "I wouldn't put that thing in my motorcycle." But the engine had performed well in tests, and it was waiting in the plane on Tuesday morning, October 14, 1947. A few days short of four months earlier, possession of the world's official air-speed record had been returned to the United States. A P-80R, a Lockheed turbojet, had achieved a speed of 623.8 mph over a measured course of 1.86 miles at Murdoc Dry Lake, California.

On that Tuesday morning in October the air was cool at Murdoc Field. The locale was an altogether benighted clump of huts in the high Mojave Desert, surrounded by a series of huge, dry lake beds and assorted Joshua trees. By the end of the morning, one more gate on the way to the stars would be opened. Out on the field a dozen men surrounded a huge silver Super Fortress. Only half visible in the belly of the B-29 was a yellow-orange plane with a nose like a long beak, slim and straight, like a hummingbird's. That beak was designed to smash through the air above the field at a speed beyond that of sound.

Fueling of the X-1 had begun a little after 5:30 A.M. The sound barrier was a couple of hours away. As the sky brightened, the yellow rocket was circled by a thick band of frost in which violet and blue vapors crept and gathered. The liquid-oxygen vapor was heavier than air, and it settled onto the concrete runway and undulated along like some primordial fog.

The supercold LO_2 poured from hoses into the belly of the plane and boiled off through vents, and the little vehicle shrieked and screamed as the metal contracted from the cold. The vents issued a snow-blue smoke. LO_2, a gas compressed into a liquid, has a natural tendency to convert into a gas again. Part of it vaporizes and expands aganst fuel-tank walls, and the vapor is shoved back against more bubbling liquid trying to become gas. The din resulting from contracting metal, boiling vapors, and agitated liquid rang out over the runway. Being fueled, the X-1 sounded and looked like some mad yellow tea-kettle with a needle nose.

The sun came up over the peak of Lonely Mountain shortly before 7 A.M. The pilot, Captain Chuck Yeager, who would be twenty-five years old in four months, arrived on a motorcycle. A half hour later, the B-29 with the yellow rocket plane slung beneath it took off. At 8000 feet, Yeager climbed into the X-1. The helmet he wore closely resembled a nylon sack held almost skin tight by strong laces. The plan was for the little ship to exit the bomb bay of the B-29 at a prearranged altitude. It would then fire its rocket engines and roar on up into the atmosphere breaking the sound barrier on the way.

At one minute to drop time, the instruments in the X-1 activated. They would tell the waiting engineers and scientists what was happening inside the plane or what *had* happened, should it disintegrate, as some predicted. On time, the X-1 dropped into the thin, cold air. Yeager hit four switches, and in quick succession four giant slams from the rockets hit him in the back. He pointed the needle beak up and the ship roared into the sky accompanied by a thin, high-pitched whistle. Conical shock waves of pale yellow cut the length of the X-1's exhaust trail. At 40,000 feet, he was traveling at 90 percent of the speed of sound. He cut off two of the rocket tubes and was still climbing steadily and quickly. Two thousand feet higher, at 92 percent of Mach 1, he was jolted and violently buffeted. His teeth chattered with the shock. Still on two rocket tubes, 8 miles into the sky and heading toward Mach 0.95, he cut the third tube back in and the instruments hung at 0.98. He was on the point of cutting in the fourth tube when the needle on his meter dipped and then went off the scale. Would the X-1 disintegrate under him?

Instead of breaking apart, the ship dropped into smooth, calm flight, in silence. All sounds were now behind him. Exactly fourteen minutes after he was dropped from the B-29, Charles Yeager landed at Murdoc. Yeager was later to remark, "It was just like any other flight except it was at 760 mph."

There were now few speed barriers ahead: the speed needed to orbit the earth; the speed needed to leave for the moon; and the velocity required for escape from the solar system. Beyond was the much-discussed speed of light and the looming spectre of Einstein.

Chuck Yeager was a long way from the moon in the X-1, but the craft had been powered by the right type of engine. While rocket engineers spent their time developing the experiments that would eventually lead them to design and build a man-carrying spacecraft, the test pilots used rocket planes to test the conditions of ultra-high-speed flight. They were waiting for the day when a man would sit at the very top of a gigantic rocket and be flung above the earth so fast that he would become a temporary "moon" in orbit around it.

Meanwhile, there were those who said it couldn't be done. Among many editorial opinions expressed in newspapers the year after Chuck Yeager's flight was one which stated with foresight, "Our candid opinion is that all talk of going to the Moon . . . is sheer balderdash—in fact, just moonshine." As late as 1956, a well-known astronomer who should have known better remarked that space travel was "utter bilge."

Yeager took the X-1 to a speed of 967 mph in 1948. And the little plane did what it was originally designed to do, as well: it took off from a runway, became airborne after 2300 feet, and was flying at 23,000 feet 100 seconds later. By 1950, the first rocket plane had been turned over to the National Air Museum in Washington, D.C. The records kept heading upward. The Douglas Skyrocket achieved 1238 mph on August 15, 1951, a little more than forty years after Barney Oldfield made headlines with his run across the frozen lake.

The X-15 was the last rocket plane to fly before a man climbed into a capsule aimed at achieving orbit. After preliminary flights in 1959, it accelerated to higher and higher speeds: 4000 mph in 1961, 4105 mph a year later. On July 17, 1962, it flew to the threshold of space at an altitude of 314,750 feet. But by then a man had already orbited the earth.

The difficulty in achieving manned orbital flight was in getting a man-rated rocket with enough propellant and thrust to reach the proper velocity. At an acceleration of three times the force of gravity (3g) it would take 9½ minutes to reach orbital velocity, and no rocket was available which could thrust for that long. At 10g the time to orbit was only 2½ minutes, but it was believed that a limit of 3g–6g was all the human body could stand. The fact that higher accelerations permitted lower thrust times did not help the situation, because even those shorter thrust times were difficult to achieve with the available rocket technology.

The human factor was also critical. It was one thing to drop from the belly of a B-29 and, reasonably slowly, to accelerate to 1000 mph. It was quite another to feel the slam of giant rockets which would kick the speed from zero at launch to 17,000 mph in only a few minutes.

The g limits had been based on the experience of fighter pilots who had thrown their planes into sharp turns and fast pullouts. At 6g, these pilots blacked out. The upper limit was established at 4g, and that for only brief intervals. No one had ever been through a sustained period of several minutes at that figure. Early tests tended to confirm this problem: at 3g, people complained of time-sense distortion and other effects.

Volunteers pushed ahead anyway and found, to their surprise, that 4g was more tolerable than 3g. They went on through 5g and 6g and all the way to 10g in tests using a centrifuge. Hardly any of the men blacked out, and one particularly hardy specimen survived 17g for a full minute without unconsciousness. The reason behind the medical "g fallacy" was quickly discovered: airplane pilots sat upright, whereas in the centrifuge (and later in spacecraft) the men were lying down. In the former case, the acceleration acted in the direction of the spine; in the latter, it operated at right angles, and the effects were less severe.

By April, 1958, the Air Force had drawn up plans for orbiting a man. The final phase of the proposal showed a landing on the moon. At the same time, the Army wanted to send a man up to an altitude of 150 miles in Project Adam, using a Redstone missile.

In October, 1958, NASA was established, and the Mercury program began as a civilian effort. Mercury test capsules were launched as early as the following August. The program was extremely cautious, with numerous test flights of equipment; several were planned with chimpanzees aboard. A final test flight on March 24, 1961, labeled MR-BD, proved that the Mercury program was ready at last to send a man into suborbital flight.

As dawn came to Tyura Tam near the Aral Sea, on April 12, 1961, a bushy-browed, twenty-seven-year-old Soviet Air Force major named Yuri Alexseyevich Gagarin strapped on a tightly laced gravity suit and walked out to a launch pad. The first manned Mercury launch was three weeks away. Gagarin, five feet three inches tall, was completely dwarfed by the 1,300,000-pound-thrust rocket pointing toward the morning sky. It is not recorded that he reflected on Wan-Hoo, or upon Daimler's engine, or upon Chuck Yeager, whom he knew about but had never met. He was not fond of horses, so it was not likely he thought about them, either.

Gagarin took an elevator to the top of a steel gantry, where he could see the flat plain and the launch pad stretched out below. On the nose of the rocket sat a five-ton spacecraft.

He nestled into a form-fitting couch and was strapped down. The spacecraft was called a *Vostok,* the Russian word for "east." He would be launched eastward over the surface of the earth; and, when the end of the trajectory was reached, he would be traveling at 17,000 mph in outer space. For a short time he would establish an orbit around his home planet, as Sir Isaac Newton had theorized more than two hundred years before.

Gagarin's wife and two sons in Moscow had no idea what he was doing. His parents, who lived on a collective farm in central Russia, a hundred miles from Moscow, thought he was just an ordinary officer on regular duty in the Soviet Air Force.

At 9:07 A.M. the engines of *Vostok* ignited with a complete burn, and Gagarin was slammed back against his couch. In a few minutes he shot into orbit at an altitude of 188 miles. In all the billions of hours that mankind had been on the earth, no man had ever before left it so far behind nor traveled around it so quickly.

"You can see a beautiful transition from the bright surface of the earth to the completely dark sky in which stars are visible," Gagarin said. "This transition is very subtle. It is as though a film ringed the earth. It is of a delicate blue color. The change from blue to dark is very gradual and lovely." He ended his description by saying, "It is difficult to render into words. . . ." Gagarin returned from space at 10:55 A.M., after staying in orbit one hour and twenty-three minutes. Had it not taken place practically in view of the whole world, the event would not have been believed except by a few dreamers, there being nothing in the history of the human race with which to compare it.

Not until early the next year did an American go into orbit, but it was clear to those who worked in the space programs of both countries that space travel had arrived at last. In the American space program, at least, there was the determination to land a man on the moon. In doing so, a new speed threshold would have to be reached. Few doubted by the time of Gagarin's flight that it would be done. The main question was physiological: could a man or men survive for more than a week in space, exposed to cosmic rays and seven days of zero-g environment? The speed necessary was what the Russians called the "second cosmic velocity." It was the minimum initial velocity needed to go from near the earth's surface to an escape from the earth's gravitational field. The first cosmic velocity, the minimum initial push needed to become a satellite of earth, had been achieved by Gagarin's flight.

Neither of these two speeds had really been a problem except in relation to a human being aboard a spacecraft. The *Sputnik* launch in 1957 had accomplished the first cosmic velocity, and *Luna 1,* on January 2, 1959, had accomplished the second. The American space program did the same with *Explorer 1,* in 1958, and *Pioneer 4,* in 1959.

The first manned assault on the second cosmic velocity was made on December 21, 1968, with *Apollo 8.* Three astronauts became the first men to orbit the moon; and, at the same time, the first astronauts to go faster than the escape velocity from the earth's gravitational pull. They raised the speed record from about 17,400 mph in near-earth orbit to more than 24,200 mph. There were then only two thresholds left to be reached: the velocity of escape from the solar system—technically, interstellar flight—and the speed of light, for genuine interstellar exploration.

On December 3, 1973, 24 seconds after 9:30 P.M. PDT, a small deep-space exploration vehicle called *Pioneer 10* made the first incursion into the system of debris, moons, radiation belts, and rings around the planet Jupiter. It was unmanned, weighed 570 pounds, with a scientific package that weighed only 65 pounds. It was only one minute off schedule after traveling more than 500 million miles.

On November 26, it plowed into Jupiter's bow shock wave; and six hours prior to planetary encounter it ran into an almost lethal dose of intense radiation. When it finally reached the planet, it raced across the cloud tops at 80,000 mph, the fastest any manmade object had ever traveled.

After the encounter with Jupiter, *Pioneer 10* sped outward from the solar system carrying a plaque showing where it had come from and who had made it—just in case anyone should find it someday.

Pioneer 10 was a state-of-the-art vehicle when it was launched in 1972. It illustrates where we are in relation to exploring the stars: our best 1972 effort was a tiny vehicle which would take more than 100,000 years to travel the distance to the *nearest* star. Had it been specifically designed as a starprobe, there are techniques which would have cut the travel time in half. But even 50,000 or 60,000 years is not an acceptable length of time for a mission.

Starflight involves the traditional relationship of all travel: distance, speed, and time. At 60 mph, Mars is 100 years away; Jupiter, 740 years; the edge of the solar system at Pluto's orbit, 6800 years; and the nearest star, 11 million years. At the speed of the X-1, Mars is 6 years away, Jupiter 27; but it would take 1½ million years to reach the nearest star. Even at Yuri Gagarin's orbital speed, the nearest star is 150,000 years away.

YOU CAN'T GET THERE FROM HERE

"Overhead and in the far distance are the lights in the sky that are stars. The stars they tell us we can never reach because they are too far away. They lie; we'll get there. If rockets won't take us, something *will."*

FREDRIC BROWN, *The Lights in the Sky Are Stars* (1953)

Crossing the ocean faster than the Concorde does would only require more powerful jet engines to double or triple the speed. A trip from New York to London might take an hour and a half by modified jet Concorde.

If we chose to convert the trip to a ballistic path with a launch, coast, and landing using rocket engines, the trip could be measured in minutes. The first investigations into rocket-powered ballistic-glide commercial flight were undertaken by Hsue-Shen Tsien in the late 1940s. There is little in the proposition that is not within the current technology of space and aircraft design, and the distance we are dealing with is little more than 3000 miles.

We have already landed on the moon, a trip of a quarter of a million miles; and *Pioneer 10* is now more than 2 billion miles out in space. Are the stars so much farther away that starflight is a near impossibility? The answer is that they are so far away that a new way of thinking about distances is necessary before we can talk about them in a meaningful way.

The discovery of stellar distances comes from the oldest branch of astronomy, called "astrometry," which deals with the space-time relationships of objects in the sky and with the measurement of small angular distances. It is a discipline that has been pursued for more than 2000 years, and one of its earliest discoveries was that of "precession."

First recorded by Hipparchus in 125 B.C., precession is the slow wobbling of the earth on its axis, like a spinning top which is slowing; the north pole of the earth describes a circle in the sky, and that part of the sky toward which the earth's north pole points slowly changes. The effect is caused by the gravitational pull of the moon and sun on the earth's small equatorial bulge. The earth is not perfectly round, and the diameter at the equator is slightly larger than the diameter at the poles; the difference is about 26 miles. The *cycle of precession,* the time it takes for the earth to trace one complete circle in the sky with its north pole, is 26,000 years.

The North Star, Polaris, was not the north star during the height of the Egyptian dynasties; and in 12,000 years, the pole star will be Vega in the constellation Lyra. Because precession shifts a star's apparent coordinates in the sky by 50 seconds of arc each year, star catalogues are listed by epoch (Epoch 1900, 1950, 2000, etc.); thus, a star's position for a given year can be calculated from the catalogues.

Precession also affects the positioning of the equinoxes against the background stars as seen from the earth. This precession of the equinoxes causes a movement of the first day of northern spring (the vernal equinox) among the constellations, although it is always on March 21. Two thousand years ago, the vernal equinox lay in the constellation Aries; it is still often called the "first point of Aries". It has now moved, because of precession of the equinoxes, to the constellation Pisces. It will lie in Aquarius in about 600 years, so the much-hyped "age of Aquarius" is still six centuries in the future.

In 1610, Galileo, with his tiny telescope, was able to see the moon as a solid body complete with craters, huge mountains, and what he thought were seas. Even his primitive telescope magnified the moon so that it appeared as if it were seen by the naked eye from a distance of 7000 miles. When he turned it on the planets, he saw Mars as a round globe, Jupiter as a great disk attended by four small moons, and Saturn as an object with "handles." His telescope was not powerful enough to show the rings of Saturn for what they were.

The magnification of the moon and planets by the telescope proved that they were relatively close to the earth. But when Galileo turned his instrument on the stars, he saw, in effect, nothing more than he could see without it. The stars must be magnified, he knew, just as the moon and planets had been. It meant that the stars were much farther away than the most distant planet known in his time, Saturn. He had no good guess on how distant the stars really were.

The distances of the stars remained a mystery until

1718, when Edmund Halley found some bright stars which were relatively displaced from the positions that had been recorded by Ptolemy in the second century A.D. The displacement was more than could be accounted for by precession. This discovery of what is called "proper motion" renewed interest in measuring small angular displacements of stars. Stellar positions were measured and then catalogued, taking into account precession and proper motion, if any (not all stars have *observable* proper motion, even in modern telescopes). Not long afterward, stellar aberration, an effect due to the finite velocity of light, was observed and explained by James Bradley. The finite velocity of light had been noted by Olaus Römer in 1675. Bradley also observed the earth's annual orbital motion around the sun, and a few years later, about 1732, he described "nutation," a small, periodic precession with a period of 18.6 years. By this time, a star's position as well as its shift, if observable was known to be a combination of a half-dozen motions, and by 1783 William Herschel had noted "solar motion," the sun's motion relative to the stars.

The science of measuring very tiny values was about to be used to determine the distance to the stars. The method was called "trigonometric parallax." A star was plotted against the others near it. Six months later, when the earth was at the other side of its orbit, another measurement was taken. The positions of the star were corrected for other motions so that the resulting difference in position, if any, represented parallax: the apparent movement of a relatively distant object when the viewing baseline is changed.

If you close one eye and look at a telephone pole at a distance and line it up with an object even more distant, you will have a position of the pole in relation to the background. If you then use the other eye to look at the same pole, the pole will appear to "jump" against the background and move. The jump is a result of changing the baseline of your observation point from one eye to another. In astrometry the width of the earth's orbit became the baseline. All that was required in order to find the distance of a star of known parallax was to know the length of the baseline, which of course is twice the distance to the sun, a value that had been known for some time with reasonable accuracy. The measure of distance was expressed in units of the sun's distance, called the "astronomical unit," or au.

Application of this technique did not happen until 1838, when an astronomer named Bessel working at Königsberg (now Kaliningrad) succeeded in measuring the parallax of the star 61 Cygni. The distance worked out to 692,264 times as far away from the earth as the sun was. It was not known if 61 Cygni was the nearest star, one in the middle distance, or halfway across the galaxy. As later discoveries proved, it was about three times as far away as the nearest star. At the time, 61 Cygni was also the star of the largest known proper motion, and for a time it was called Bessel's Star.

The science of measuring small changes in position had proved its worth; the distance to a star was known, and there would be other distances determined in the coming years. The distance allowed astronomers to put the earth, sun, and solar system in perspective with the rest of the known universe. In 1844, the astronomer who first measured parallax and thus found the distance of a star also discovered "perturbations" in the proper motions of stars. This tiny measurement of position change led to the discovery of unseen companions of stars and would, in the twentieth century, lead to the apparent discovery of planets orbiting other stars.

The measurement of stellar distance in terms of the sun's distance resulted in inconveniently large numbers. 61 Cygni was 692,264 au. In miles, the figure was even more unwieldy: 68 trillion, or 68,000,000,000,000. A new unit was conceived called the "parsec." The word was derived from the term "parallax second" and was defined as the distance of a star whose parallax movement against the background stars was 1".0, one second of arc. The parsec was equal to 206,265 au, or 19 trillion miles. Determined by this method, the distance of 61 Cygni was 3.4 parsecs, or 3.4 pc.

The parsec has never caught on with the general public, although it is the unit of choice in much of astronomy. Perhaps the reason is that the public is more interested in things like the speed of light, which leads to the light-year. People are not so interested in a seemingly dull subject such as geometry, which gave us the parsec; but geometry is essential to astronomy, which is why astronomers favor this unit of measurement.

The velocity of light, universally referred to as *c*, can be measured by determining independently the wavelength and the frequency of a particular spectral line. The modern value for the speed of light is 186,282.397 miles per second (or 299,792.458 kilometers per second); the value is the speed in a vacuum. Light travels more slowly in denser mediums, such as glass. The speed of light in a vacuum is the same for all wavelengths. If the value for the speed of light is measured in miles per hour and multiplied by the number of hours in a year (or miles per second by the number of seconds), the result is the distance, in miles, that light travels in one year. This distance is about 6 trillion miles (9½ trillion kilometers), or one third of a parsec, or 63,242.01 au. The light from 61 Cygni took 11.2 years to get to the earth at the speed of light; thus, its distance was 11.2 light years (ly).

Still using manual techniques, measuring tiny angles, astronomers finally found the nearest star. It had not been found earlier because it was in the Southern Hemisphere and not as accessible to telescopes of the north where most of the work in astronomy was being done. The star was a multiple system, of which the two bright components formed the visible star seen from earth as a point of light. There was a third and smaller companion star, very faint, which was actually the closest star. The entire system was called Alpha Centauri—sometimes Rigel Kent—and the small companion was named Proxima Centauri, meaning "closest." The system was 26 trillion miles away, with a parallax of 0".765. It was also 1.3 parsecs, or 4.3 ly. Proxima was about a trillion miles, a tenth of a light year, closer to the earth than the two main stars.

The second-closest star was Barnard's Star, at 5.9 light years; and Sirius, one of the brightest stars as seen from earth, was 2.6 parsecs, or 8.6 light years. Sirius was the sixth-closest star. Arcturus was sitting at 36 light years, and within a span of nearly 200 trillion miles there were only 75 stars.

As astronomers improved distance measurements, they found the visual method of checking parallax had a limit. The photographic plate replaced the astronomer's eye, and establishing distances was improved, but there were still stars that showed no measurable parallax. There was a practical limit for direct parallax measurements of about 35 parsecs, 100 light years. Beyond that the distances were estimated by analyzing starlight. With some uncertainty over the figures, the results were surprising.

There were perhaps 10,000 stellar systems within 100 light years, the limit of parallax measurements. Using analysis of starlight, astronomers found that the diameter of the Milky Way galaxy was 100,000 light years. It was shaped like an enormous lens, 15,000 light years thick. It was estimated the galaxy contained 200 *billion* stars. Our solar system was located one third of the way from one edge, 30,000 light years from the center. Light from stars near the galactic core started its journey long before mankind emerged from caves to become an agricultural species living in primitive lake villages.

In a further stretch of the mind, astronomers calculated the distance to the nearest neighbor galaxy, Andromeda (two small galaxies are closer, but they are satellite galaxies of our Milky Way, called the Magellanic Clouds). Andromeda was found to be 2.2 million light years away, twice as large as our galaxy, with an estimated 300 billion stars. Along with the Milky Way and several others, the Andromeda galaxy made up what astronomers called the "local group."

Beyond Andromeda, in twentieth-century telescopes and photographic plates, they saw millions upon millions of light years populated with pulsars, quasars, galaxies, and more stars without number. At the greatest reach of the 200-inch Mount Palomar telescope, the largest in the world, they saw clusters, not of stars but of galaxies—galaxies, someone wrote, "like grains of sand." They were looking at distances of at least 2 billion light years.

Once the vast distances were known and digested, those who dreamed of starflight were subdued. Flashing across the deep void of our galaxy to Andromeda seemed impossible. Even Rigel, a comparatively close star in the Milky Way, was more than 500 light years away. Discussion of starflight was confined to the suns of the solar neighborhood, especially to Alpha Centauri, Barnard's Star, and 61 Cygni.

Even the idea of a trip to Alpha Centauri was hard to envision. By galactic comparison the star system was nearby—slightly over half the average distance between stars. Light takes only 1.25 seconds to reach the moon, but *Apollo 11* took four days; at light-speed, Mars is reached in less than half an hour, but the unmanned Viking spacecraft took a year to arrive; light can cross the solar system within the orbit of Pluto in only 11 hours, but Alpha Centauri is more than 4 years away. A round trip to Rigel at the speed of light would take 1000 years to complete.

To get to the center of the galaxy would be a one-way trip of 30,000 years, at least from the viewpoint of those people who might watch the ship take off from earth. There are other time frames for those who might be aboard such a ship, as will be apparent later on. The trip to Andromeda at light-speed would take 2½ million years with earth as a launch point.

The immense distances meant that incredible speeds would have to be attainable for mankind to attempt starflight. *Pioneer 10,* with its more-than-100,000-year mission time to the nearest star, is an impossible concept as a starship. At ten times the speed developed by *Pioneer 10,* the nearest star would still be 10,000 years away in mission time. In real terms, the speeds necessary for starflight would have to be measured in tens of thousands of times the speed of *Pioneer 10* if a round trip were to be attempted within the lifetime of a human crew. And those speeds present certain problems. Since we cannot alter the distances to the stars, other variables in the distance/speed/trip-time relationship have to be investigated.

The distances to the stars actually do change, but very slowly. The system of Alpha Centauri is moving in the direction of the sun, and in 28,000 years it will only be 3.1 light years away. The second-closest star, Barnard's, is also moving in our direction, and at 67 miles per second. By A.D. 11,800, Barnard's star will be at 3.75 light years and will be the *closest* star. This star has such a large proper motion that in 180 years the distance it moves in the sky is equal to the apparent diameter of the moon. If all the stars in the sky were moving as fast, and at random, the forms of the constellations would be altered appreciably in the length of our lifetimes.

It might be concluded that if the ability of mankind to reach greater and greater thresholds of speed increased 1000 times in 70 years—from 999 in 1904 to *Pioneer 10*—then a further interval of 70 years of similar technological progress would result in available speeds of 100 million miles per hour. If this were true, then trips to Alpha

Centauri in the year 2043 would take 30 years. Allowing for time to slow down at the star system and for a period of exploration, a round trip might just be within the lifetime of a future crew, assuming that medical and biological progress paralleled the technology of spaceflight.

An additional passage of 70 years at the same rate of progress would result in speeds measured in billions of miles per hour in the year 2113. The trip to the nearest star would be a short round trip. It is easy to imagine there would be volunteers for such a "short" flight of magnificent exploration. It is equally possible that by 2113, the means could be found to sustain them that long in space aboard some kind of ship. Even now, our deep-space unmanned spacecraft have lasted 7 years in flight. *Viking Lander 1* on Mars is expected to keep functioning for 15 years.

The problem with this sort of straight-line reasoning from the history of technology is that it doesn't necessarily hold true. Technological progress does not always move in a straight line, or even in a wildly logarithmic curve. There are reasons for the rapid advance of technology in this century, and those reasons do not necessarily dictate that the next century will progress at a similar rate. That a society is theoretically able to accomplish something does not mean it will do so; a technology must have an application that will justify its continuing development.

The energy in the wood or coal that was necessary to drive a transcontinental railroad in 1870 is similar *in magnitude* to the energy needed to propel its locomotive to orbit. Our forefathers in the late nineteenth century ran their trains on rails, of course, not having the foggiest idea of how to send a locomotive into orbit. They would have seen little application for such "technology," though they theoretically had the energy to do it.

That the locomotive gave way to other forms of transportation, ones that increased speeds dramatically, was a happy coincidence of several factors. The most important factor was a change in propulsion methods and fuels: steam to gasoline, to liquid hydrogen and liquid oxygen; locomotive to automobile to airplane to rocket ship. It did not occur as a natural event nor as a consequence of steam technology and the passage of time.

Nor did the increase in speeds derive from sudden genius. Though it is true that Tsiolkovsky was the first to propose rocket technology on paper and to suggest the use of liquid hydrogen and liquid oxygen as propellants, Hermann Oberth independently published quite similar calculations not much later in Germany. Liquid propellants were based on the principles of chemical thermodynamics, the foundations of which had been laid by 1880. Also, though Daimler was the first to perfect the gasoline engine, there were many engineers at the same time who were independently investigating the construction of an internal-combustion gasoline engine.

Steam technology developed rapidly because it had obvious commercial applications: ships, locomotives, personal transportation. It clearly allowed greater speed, the transportation of greater loads for less cost, and was both profitable and practical. The gasoline engine, which in automobiles and airplanes led to greater and greater speed records, was obviously suited to commercial application. In an era of economic aggressiveness their development was assured.

Although the commercial aspects of rocket technology were not new (rockets were used to propel lifesaving apparatuses between ships, for example, and during the '20s and '30s experiments were performed in Germany using rockets to deliver the mail), it was the military applicability of the rocket which led to its rapid development.

President Dwight D. Eisenhower correctly defined the main force behind the fantastic development of speed over the last century: he called it the "military-industrial complex." If even the most tenuous relationship to historical technological growth is to be maintained, flight to the stars must eventually have some commercial or military application, or the technologies needed for flight to the stars must be developed for reasons of economic or military value and then be applied to "pure exploration."

A further and more devastating argument against the idea of straight-line technological progression which would end up allowing us to fly to the nearest star in a few minutes several hundred years from now involves the speeds we can achieve. If, as projected, in the year 2113 we would have exceeded a billion miles per hour, that would be faster than the speed of light, and physicists remain convinced that the speed of light cannot be exceeded.

For more than two hundred years, the world of astronomers and physicists was dominated by the figure of Sir Isaac Newton. While still a very young man, Newton had in a single summer produced three theories to serve as solid foundations for studying the universe. The formulation of the Law of Universal Gravitation was the greatest of the three. He constructed a logical and mathematical synthesis of the physical world and the universe. It was a comforting synthesis in which all objects—men, birds, planets, stars, particles, and the stuff of which light consisted—moved in accordance with the same mathematical laws. This system became the basis of nineteenth- and early-twentieth-century physics.

Newton used two words whose definitions underlay his whole system. The two words were "space" and "time." He described a universe with "absolute, true, and mathematical time" flowing equably without relation to anything external, and "absolute space, in its own nature, without relation to anything external." It was a very mechanical and straightforward universe, where events did not confound scientists by doing things they shouldn't.

But, as Sir Arthur Stanley Eddington was to remark, "The Newtonian framework, as was natural after 250 years, had been found too crude to accommodate the new observational knowledge which was being acquired. In default of a better framework, it was still being used, but definitions were strained to purposes for which they were never intended. We were still in the position of a librarian whose books were still being arranged according to a subject scheme drawn up a hundred years ago, trying to find the right place for books on Hollywood, the Air Force, and detective novels."

Albert Einstein was a young man who, at sixteen, had already found something rotten in the state of Newtonian physics. He once remarked, "I was supposed to choose a practical profession, but this was simply unbearable to me." In a letter to his uncle in 1895, the young Einstein enclosed a five-page essay written in a sloping and spidery Gothic script. He had discovered a paradox by reflecting on what would happen if he could follow a beam of light at the speed of light. This paradox and other questions about the Newtonian universe weighed on Einstein's mind for a decade.

The Special Theory of Relativity gave Einstein his unique position in the history of scientific thought, and it is this theory which poses so many problems for starflight. The Special Theory was outlined in the third paper Einstein wrote for the *Annalen der Physik* in the summer of 1905. From the time he grasped it fully until the time it was published, only six weeks had elapsed. The paper was titled "On the Electrodynamics of Moving Bodies," and it was not the usual academic-journal entry: Einstein held forth for 9,000 words without one source reference or footnote.

The Special Theory rested on the principle that the velocity of light is the same for all observers, no matter how they move relative to each other. That, of course, sounded basically paradoxical, and in the end it forced scientists to abandon the familiar Newtonian notions of space and time. Special Theory said time runs at different rates for different observers moving at different velocities.

The constancy of the speed of light for all observers means that two spaceships approaching each other at the speed of light still only register a closing velocity of *c*, the speed of light, not 2*c*, as "common sense" would dictate. Two spaceships receding from each other at the speed of light show a velocity of recession of *c*, not 0.5*c* or 2*c*. Not only was the speed of light the same for all observers regardless of how they moved relative to each other, but Special Theory indicated the speed of light could not even be *reached*, or *exceeded*. To exceed the speed of light, a spaceship would, however momentarily, have to *reach c*. (For example, an automobile going from 0 to 100 must, however briefly, pass through all the speeds in between. It cannot go from 51 to 56 mph instantaneously and thus bypass the national speed limit while exceeding it.)

The London *Times* in 1905 said the Special Theory defied common sense. The newspaper was correct, but failed to point out that the Creator of the Universe, if there was one, might never have heard of "common sense." In the world of Newtonian physics a force applied to an object results in speed. ("For every action there is an equal and opposite reaction" is Newton's Third Law, the principle by which all rockets work.) If the force is doubled, so is the speed, assuming the events take place in a vacuum or on paper. According to Newton, this event should keep going indefinitely until an infinite speed is reached, always assuming an infinite force is available. "Common sense" would also indicate that this is true. It was certainly true after 1947, when aircraft went faster than sound. More powerful rockets raised the speed from Mach 1 to Mach 2, then Mach 4 and higher. Gagarin was going at Mach 25 when he orbited the earth.

But all of these speeds were only a tiny fraction of the speed of light; there was no reason to exchange Newton's views for those of Einstein. When very high speeds are considered, however, speeds approaching *c*, Newton's laws do not apply.

The reason for this is simple: if we cause an object to go from rest to a certain speed and then give it an additional push to add more speed, part of the second push goes into increasing the *mass*, and part of it goes into increasing the speed. At low speeds in relation to that of light, the amount which goes into increasing the mass is very small, so two times the push seems to result in twice the speed. But at higher values, more and more of the push goes into increasing the mass and less and less goes into increasing the speed. At speeds significantly near light, the mass of the object increases very greatly and the speed is increased only slightly.

The result of all this is that near *c* almost all of the push goes into increasing the mass and practically none of it into increasing the speed: the Sisyphus Effect.* If an infinite amount of push is put into an object, it simply becomes an infinite mass, and the speed that results is 100 percent of the speed of light. Since a spaceship cannot have an infinite mass (it would equal that of the universe), and there is no infinite power available (the total power of the universe would be needed), the speed of light "cannot be reached."

The Special Theory had several messages for starflight, almost none of them encouraging. The speed of light could not be reached or exceeded. This meant that no conceivable propulsion system could cause a starship to reach even the nearest star in less than 4.3 years, its light-distance. Allowing time to accelerate to near *c* and time to decelerate to the target would add another year at a minimum. No one-way trip to Alpha Centauri was

*Sisyphus was a mythical Corinthian king whose job in Hades was eternally to roll a stone up a hill only to have it roll back again.

going to be shorter than 5.3 years and a round trip with exploration time at the target star would occupy the better part of a dozen years.

The trip times were, however, *relative.* From the viewpoint of a crew aboard a starship at the speed of light, the time would be shorter (time does not everywhere proceed at the same rate, Einstein said). The rate of time flow depends on the speed of the clock measuring it. At near c, the clock rate is near zero relative to a stationary point (which in Einstein's view could not exist, anyway, as all points are moving in some way relative to one another). At $v/c = 0.999999$, what seems like 5 years to a starship astronaut traveling at that speed would be 3535.53 years to an observer at rest. The clock aboard a starship going at $v/c = 0.5$ (half light-speed) reads 0.867 hours for every hour that has passed for an observer at rest: 52 minutes elapse on board for every hour elapsed for the observer at rest. This is the "time-dilation effect."

Time-dilation effects tend to cancel out some of the problem in exceeding the speed of light. A thirty-year-old astronaut could leave earth for a 14-light-year journey at 99.9 percent of the speed of light and arrive at his target having aged only 2 years. His earthly counterpart at the point of launch, however, would have aged 14 years. What time dilation means to starflight is that all distant trips are essentially one way unless the speed of light can be exceeded. A long starflight at light-speed would have an astronaut returning thousands of years in the earth's future, while only a few years would have elapsed for him.

Until the twentieth century, there were no objects that traveled fast enough to cause anyone to worry about Newtonian physics. But with the discovery of subatomic particles, things began to behave according to Einstein's predictions and not Newton's. The Special Theory and its provisions, its postulates, have been experimentally checked thousands of times over the past three quarters of a century. So far, there has been no solid reason to doubt that the Special Theory is correct, that Einstein was accurate, and that the universe behaves according to his theory. There is as yet no reason to think the speed of light will ever be exceeded, and it remains a barrier to starflight in terms of acceptable mission times.

The predictions of the Special Theory were well known long before the technology of the liquid-fuel rocket became available. It was obvious to even the earliest speculators upon starflight that most of the relativistic effects predicted by Einstein could be avoided by traveling at half the speed of light or less. Unfortunately, that meant a mission time of 22 years for a round trip to Alpha Centauri; the time dilation at that speed was negligible. At 0.1*c*, still a fantastic speed, the relavitistic effects were nil, but Alpha Centauri was more than a hundred years away. Reaching more distant stars at this speed was out of the question.

Obviously it is *possible* to travel to the stars, despite the Special Theory and Einstein. The question for early speculators about starflight, and for us today, is whether it is possible to achieve spaceship speeds great enough to bring trip time within the limit of reasonable human life expectancy. This involves the propulsion available to us, and what might become available in the future.

RUBBING TWO STICKS TOGETHER

"You can't reach the stars just by rubbing two sticks together."

ANONYMOUS

The most recognizable feature of the Space Age is the thundering sound of a rocket taking off. Away it goes, shaking the ground and ripping the air until it becomes too small to see or is obscured by clouds. Unfortunately, it uses most of its available thrust—and fuel—in getting off the ground and a hundred miles into orbit. A rocket would be much more efficient if it could be built and launched in space, where the earth's gravity is less of a factor. The famous phrase "Once you're in orbit, you're halfway to anywhere" is essentially true: an orbiting spaceship has 70 percent of the velocity needed for complete escape of the earth's gravity, and it's a velocity won the hard way in terms of fuel.

Travel to the stars requires such fantastic fuel loads to cross the great distances that no starship powered by any means of which we can conceive would be able to do it from the surface of the earth. For the present, and for the foreseeable future, the idea of taking off from starport New York and landing in starport Nova Terra, around Alpha Centauri, is confined to fiction.

Any starship we can envision will be built and launched from an orbit within the solar system so that all of the propulsion energy can go into escaping the solar system and making as quick a trip as possible to a star. One consequence of this is that a starship project must ultimately depend on an expanding space development into high earth orbits, lunar orbits, and positions around the other planets; another is that a starship would lose the other recognizable feature of rockets: it would not need to be aerodynamic in any way. There is no air in space, and no air resistance which increases with speed. The slim projectile shape of rockets as we know them would give way to an engineering design that would be functional for the medium in which it travels. The Apollo lunar-landing vehicles, the LEMs, are a good example of a rather ugly but functional design for use where air resistance does not exist.

The liquid-fuel chemical rocket has been around since a man named K. E. Tsiolkovsky, from an obscure part of Kaluga Province in Russia, invented it in 1896. In 1903, he published a paper entitled "Exploration of Space with Reactive Devices." In the article he made five simple statements: (1) rocket travel is possible; (2) only with rocket propulsion is it possible, since only a rocket can work in outer space, where there is no air; (3) gunpowder rockets are not powerful enough; (4) certain liquids *are* powerful enough; (5) liquid hydrogen would be a good rocket fuel, and liquid oxygen would be a good oxidizer for it. Tsiolkovsky's rockets existed only on paper; and, writing in what was to the Western world an obscure language, he was not widely read.

His fifth statement could not have been made with authority much earlier. It is a perfect example of how science and technology rely closely on what has gone before. One discovery leads to another and to another, but often none of the results would have been possible without one particular event of *primary* significance. Scientists had been trying since 1823 to liquefy gases by cooling and compressing them. Chlorine was the first to be made liquid successfully; it was followed by ammonia, carbon dioxide, and others. By 1870 a few gases were still not available as liquids. Among them were oxygen, hydrogen, and nitrogen, in addition to fluorine, which had not yet been isolated, and the rare gases that had not yet been discovered.

In Krakow in 1883, a scientist named Wroblewski and a colleague, Olszewski, managed to liquefy oxygen. By 1891, it was available in experimental quantities and by 1895 in large amounts. Hydrogen was first made liquid in 1898, by Sir James Dewar, the inventor of, among other things, the vacuum bottle, or Thermos. Without the work of Wroblewski and Dewar, Tsiolkovsky would not have been able to speculate so accurately on spaceflight's dependence on liquid fuels.

Robert H. Goddard, who had never heard of the Russian theorist, reached almost the same conclusions. Goddard's rockets used liquid oxygen (referred to as LO_2) and gasoline. The German V-2 used alcohol and LO_2. But many subsequent chemical rockets, from the one that launched the world's first artificial satellite to the space shuttle, use liquid hydrogen (LH_2) and LO_2 stages exactly as Tsiolkovsky had predicted. They differ from the rockets of Robert Goddard only in sophistication and development. They have liquid fuel in one tank and a liquid oxidizer in another. The two are brought together in a combustion chamber and ignited.

While it is true that there is such a thing as perfection of design, the chemical rocket needs improvement. The Apollo-Saturn moonship was capable of sending "only" 150 tons to earth orbit, 50 tons to the moon. Had it been launched from an earth orbit, it could have landed more payload on the moon. But even with that advantage, it would have been a slow boat to Mars. As a starship, it would be worthless. Had it been aimed at the nearest star and launched from earth orbit, it would have taken 70,000 years to get there.

In the early days of rockets, experimenters tried anything that would pour or burn and some things that would not. The quest for more power was the realm of the chemical-rocket engineers, and they were an amazing and suicidal group. LO_2 and gasoline was tried, in continuing the experiments of Goddard. LO_2 and liquid methane was used. So was gasoline and nitrogen tetroxide. The substances the engineers were playing around with were temperamental in nature and, of course, frequently dangerous.

Esnault-Pelterie tried tetranitromethane as an oxidizer and promptly blew off four fingers. In Leningrad, Glushko did the same, omitting the fingers part of the experiment. Sänger went through a long list of fuels, from hydrogen through pure carbon used with oxygen and other oxidizers.

An Italian, Luigi Crocco, tried "monopropellants": fuel and oxidizer in one liquid. With crossed fingers, he tried nitroglycerine and almost joined Wan-Hoo as a casualty of rocket history. His mixture was slightly tranquilized by having added 30 percent methyl alcohol, but that wouldn't have helped much if the molecules had gone the wrong way. He also tried nitromethane, a slightly less sensitive substance.

Before the V-2 had been perfected, the propellant engineers had gone through hundreds of combinations: alcohols (various kinds); saturated and unsaturated hydrocarbons; lithium methoxide; dekaborane; lithium hydride; and aluminium trimethyl in combination with oxygen; red fuming nitric acid (a thoroughly charming substance); and nitrogen tetroxide. In 1944 Bob Truax, a man of the true breed, tried a mixture of benzine and tetranitromethane, which, of course, detonated immediately, without making any attempt at becoming a new fuel.

Some of the chemicals were so corrosive that they could not be loaded into the rocket until just before firing; some gave off dense clouds of poisonous fumes such as nitrous oxide. Some were so potent that painful burns resulted from one drop.

Fluorine was considered as an oxidizer, and it was, as one pioneer said, "a holy terror to handle." A substance called "aniline" was subtle in its effect: if it wasn't washed off a man immediately, he would turn purple, then blue, and die within minutes from cyanosis. Some of the fuels had an indescribable stench, tending to remain on clothes and skin for weeks. Experimenters even tried lemon oil, and they tried nitrogen tetrafluoride plus hydrazine, the former a substance put on furniture and the latter a high-energy compound with a very low boiling point.

Immediately after World War II, the United States somewhat belatedly realized the importance of rocket research and established several groups to work on fuels. The switch was so abrupt that engineers who one day were working on reports about spark-plug fouling or flame-front propagation in cylinder-head combustion chambers were the next day working on rocket-cooling research. Some of the early groups in the U.S. began with hydrazine, diborane, and ammonia, and continued oxidizer research with hydrogen peroxide and liquid oxygen. The emergence of computers allowed researchers to evaluate fuels and oxidizers without actually having to mix the sometimes cranky substances together.

When all the experimentation was done, it was concluded that Tsiolkovsky and Goddard had been essentially correct in 1903 and 1909: liquid hydrogen was an ideal rocket fuel for obtaining high thrust per pound, and liquid oxygen made an excellent oxidizer for it.

The Saturn V rocket had a thrust of 7.5 million pounds (or 56 million horsepower at 2800 mph).* It was fueled by a kerosene-and-LO_2 first stage, with upper stages using hydrogen and oxygen in liquid form. The space shuttle is an all– LH_2-LO_2 vehicle. The moon rocket and the space shuttle are fueled by the hydrogen-oxygen mix because it produces a respectable power output and is fairly easily handled. Is there some sweet combination in the grab bag of fuel experimentation that would provide the speeds necessary to reach the stars? If there is, it must be both extremely light and extremely powerful.

Rocket performance is measured in many ways, but in the end it comes down to efficiency. Rocket efficiency is measured as "specific impulse," in seconds, usually written Isp. Though there are complex technical definitions of Isp, it can be thought of as the time in which 1 pound of propellant produces 1 pound of thrust, or the thrust that 1 pound of fuel can produce for 1 second. *Thrust* of a rocket engine depends more on design of the engine, whereas Isp depends on the energy of the fuel and the conditions under which it is used: pressure, temperature, etc.

The best kerosene-oxygen engines available have Isp's between 250 and 300 seconds. Liquid hydrogen–liquid oxygen engines have Isp's somewhere in the range of 400–450. As early as 1959, a hydrogen-fluorine engine reached almost 100 percent of theoretical performance

*7.5 million pounds' thrust = 56 million horsepower at 2800 mph (4100 feet per second) velocity. There is no direct conversion between thrust and horsepower; the velocity must be specified.

at 480 Isp. But Isp 500 or so does not translate into a practical starship engine. A hydrogen-fluorine rocket, even if enough fuel could be taken along (it can't), could not achieve 0.01 percent of the speed of light (psol); at that speed, which is 18.63 miles per second, the nearest star is still 42,000 years away. If some unforeseen breakthrough should occur, there is still no chemical possibility which would bring the nearest star closer than perhaps 20,000 years for the trip. Chemical fuel, which has served us so long in the early penetration of space, won't do the job. We must turn elsewhere for fuel for a starship.*

As in the change from propeller-driven aircraft to jets and rockets, a new type of fuel and a new type of engine are needed to have a chance at starflight. Fortunately, both are potentially available in the nuclear rocket and in a few other forms that could become available sometime in the future.

The so-called atomic rocket does not depend on the combustion process as chemical rockets do. It consists mainly of a tank holding the "reaction mass" to be expelled and an atomic reactor run at a temperature as high as possible without a meltdown. Experiments on nuclear rocket engines were done in the 1960s during Project NERVA (Nuclear Engine for Rocket Vehicle Application) using hydrogen as the reaction mass.

In operation the atomic rocket uses a pump to bring the reaction mass into a jacket surrounding the reactor. The hydrogen is heated to a gaseous state and flows through a multitude of holes in the reactor core where it is heated further. The liquid hydrogen emerges as a very hot gas and is expelled through an exhaust nozzle.

This type of nuclear rocket is called the "fission solid-core." It has several drawbacks, one being the long start-up time necessary because of the temperature difference

*The NERVA engine was limited to about Isp 850 sec.

between the cold liquid hydrogen and the hot reactor core. Also, the reactor is slow to "stop." When it is shut down, it stays physically hot for a long time, as well as radioactively hot.

On March 3, 1966, a NERVA engine was run under full power for the first time. It proved to have about twice the exhaust velocity of a hydrogen-fluorine rocket.* Although it was 70 percent more powerful then a chemical rocket on paper, the engine weight was so great that the actual benefit in speed over chemical propulsion would be in the neighborhood of 50 percent, with some uncertainty. Development of the NERVA engine was abandoned after massive cutbacks in the space program in the early 1970s, which left the project in search of a mission.

A further step in the development of nuclear fission rockets is the gas-core design. There are limitations on the performance of solid-core nuclear rockets because of restrictions on temperatures. These restrictions can be removed if the fuel is a gas instead of a solid and high temperature is used. Gas-core designs studied in the early 1970s were proposed with Isp near 2000 seconds, an improvement of four times on any foreseeable chemical fuels. Unfortunately, this is still far short of a starship-propulsion system.

A nuclear gas-core fission rocket would work for the relatively short trip to Mars—one was, in fact, proposed in 1972, with a five-man crew and a trip of 30 days—but it has yet to be developed after more than 20 years of research. Some of the best experts in nuclear rocket propulsion at Lawrence Livermore Laboratory in California do not believe, after two decades of study, that a gas-core rocket can, in fact, ever be built. Should it turn

*To launch a 1-pound payload to one quarter the speed of light with Isp 2000 sec. would require propellant of 10^{1660} pounds. The mass of the universe is about 10^{53} pounds. Obviously any such fission rocket would travel very slowly to keep the fuel weight within the bounds of mathematical possibility.

out that one can, it is still not satisfactory for starflight; unless the trip were very slow, it could not carry enough fuel for a stellar voyage.*

The efficiency of fission reactions is only 1 percent. When a fast-moving neutron, a subatomic particle with no electrical charge, strikes a large nucleus such as uranium-235, the result is a splitting of the uranium nucleus. The debris, composed mostly of neutrons and lighter nuclei, weighs less than the original nucleus. The absent mass has been converted to energy via Einstein's ubiquitous $E = mc^2$. If a sufficiently large number of other uranium nuclei are nearby, they will be shattered by the high-speed debris of the first nucleus and the process will be *critical*—self-sustaining. In uncontrolled form, the result is what devastated Nagasaki; controlled, it can power a submarine or a city or a rocket. If somehow the fission process could be improved beyond the 1-percent level, a proper starship engine might be obtainable. Improvement is possible, but the process has yet to be perfected; it is called *fusion,* often described as the ultimate energy source.

Fusion is the opposite of fission. Nuclei of light elements such as hydrogen are forced together by high-temperature impacts. Other subnuclear fragments come together at the same time, and the result is a heavy nucleus. The new one has less mass than the total of the parts that were fused to form it. Again the absent mass has departed as energy. While fusion reactions liberate only 3 to 5 times more energy per unit mass than fission, they lend themselves to the production of high-temperature plasma fireballs that can be efficiently coupled into an exhaust beam.

Fusion is the only propulsion system within our grasp

*The mass of both particles is converted to energy with a high efficiency. Annihilation of electrons by positrons yields 100-percent energy release, the result in the form of gamma rays. Annihilation of baryons by antibaryons yields a flood of pions and other short-lived particles, so this annihilation is not 100 percent efficient.

which can propel a starship at a respectable percentage of the speed of light. Even so, a fusion-powered ship would not be able to travel *near c;* a small 5-ton starship attempting to reach 0.97*c* would require the services of a small star for fuel. The process, however, is not quite as easy as it sounds. Nuclei tend to resist fusing together, and fusion has yet to be achieved in a laboratory.

The electrical charge on a nucleus is positive and it will repel another nucleus. To overcome the natural resistance to fusing, three conditions are needed: high temperature, high density, and confinement for a length of time sufficient for fusion to occur and the process to be self-sustaining. Stars are fusion reactions, but in stars the necessary conditions are arranged by the huge gravitational forces. On earth, gravity is much too weak to be a useful force in a fusion experiment.

One of the most promising efforts which may lead to fusion on earth involves a *tokamak,* a device invented in Russia in the 1950s. A tokamak is a gigantic doughnut-shaped chamber filled with a gas. Outside the chamber is a circle of electromagnets. The gas can be heated by one of a variety of methods, including radio waves operating in a fashion similar to a microwave oven. When a temperature near that required for fusion has been reached, the atoms of the superheated gas lose their electrons and become a plasma which is composed of ions and negatively charged electrons. The plasma, still superheated, is confined in the tokamak by the magnetic field generated by the electromagnets. The giant field also acts on the plasma to compress it and raise the density. In theory, in a tokamak, all three of the requirements for fusion can be achieved: temperature, density, and confinement.

The trouble is that all tokamaks built so far do not deliver as much energy from the fusing plasma as that which has gone into raising the temperature and density. Fusion, in other words, has yet to be achieved. Nevertheless, experiments with equipment at Princeton have led many scientists to believe that the point of fusion is extremely close in time. In 1980, magnetic confinement techniques at Princeton led to a plasma temperature of 148 million degrees F (82 million degrees C). A new fusion experiment at Princeton, called the Tokamak Fusion Test Reactor, is scheduled to begin operation in 1982. The likely result of the new equipment is at least a break-even point for the plasma, and that would be a large step toward eventual perfection of a fusion process.

At the same time that fusion experiments have proceeded at Princeton, other scientists have been trying different means toward the same end. Magnetic mirror confinement of a superhot gas has been tried at Lawrence Livermore, and another major direction called "inertial confinement" has been tried at the University of Rochester. The latter method involves creating a series of small explosions by compressing pinhead-size pellets of deuterium—heavy hydrogen. The compression is achieved by illuminating the pellets with high-energy laser beams or bombarding them with ion beams. There have been predictions that laser fusion or magnetic mirror confinement will result in fusion breakthroughs by the mid-1980s.

One of the keys to making the predictions about fusion for the near future has been the Magnetic Fusion Engineering Act, which was signed into law in late 1980. The act represents a dedication to finding in the fusion process an energy solution for the United States. The amount of money allocated to fusion research, $20 billion, has already been challenged as insufficient, and more money for fusion research seems a political probability.

But even when fusion is developed as a possibility for starships, all of the problems will not be solved. The most difficult problem may be that of shielding. It is generally agreed that fusion-reactor design presents severe problems of neutron activation and irradiation, which are at least as hard to handle as the side effects of fast-breeder reactors and much more so than those of light-water reactors. Another difficulty is obtaining "fuel" in sufficient quantity.

A continuous-fusion starship would use deuterium-tritium fusion for the ignition, with deuterium-deuterium fusion for added boost. D+D reactions are more difficult to light off than D+T and will require substantial future development to be anywhere near practical. The advantage, however, of the D+D type is that deuterium is common in seawater which, of course, is abundant on earth. Tritium, on the other hand, must be bred.

The third of the "big three" fusion reactions is $D + {}^{3}He$—deuterium and helium-3. It is the least desirable of the possible starship fusion fuels because it is not particularly abundant in nature except as ions in the solar wind. On earth, the supplies of helium-3 we now use are obtained at great expense by bombardment of lithium-6, which is much more common. Helium-3 is so costly on earth that the price is listed in millions of dollars per ton. The one bright hope for the future, however, is that helium-3 could be obtained from the ordinary helium which occurs in massive quantities in the atmospheres of Jupiter and Saturn.

Any of these "common" fusion reactions could be used to power a starship; and since 1964, a continuous-fusion propulsion system has existed—at least on paper—which would reach Isp 500,000 sec, 1000 times better than the best chemical rocket.

A 1972 idea by Nuckolls, Wood, and others at Lawrence Livermore Labs theorized an implosion system energized by a high-energy laser. The scheme would allow efficient burn of small pellets of hydrogen isotopes using laser technology. For those who were working on ideas for starships, the proposal was important. It led directly to what were called "a class of specific-impulse-maximizing fusion-propulsion engines"—on paper, of course. These engines used deuterium-tritium pellets which were injected by a rotating mechanical accelerator into a thrust chamber at 500 pellets each second. Each pellet, as it passed the fusion point, was struck isotopically by a laser pulse (the pulse was infinitely quick, less than 1 billionth second). The laser pulse was focused on the pellets by optical mirrors.

In 1975, Dr. Thomas Heppenheimer, in an article in the *British Interplanetary Society Journal,* showed that the specific energy of the D+T reaction indicated a limiting Isp = 2,640,000 sec. Though he did not propose that a 100-percent-efficient engine could be produced, he believed that the theoretical limit for a fusion-engine design was about Isp = 1,000,000 sec. He proposed using a very-high-energy pulsed laser fusion system. A single-stage rocket with Isp = 1,000,000 sec could achieve $0.1c$ with a mass ratio of 20, and a staged rocket using the laser-fusion engines could get much closer to the speed of light. What the advances in fusion during the 1970s and experiments in the 1980s have shown is that the interstellar spaceship *may move from the realm of paper and speculation to the stage of preliminary design* sometime early in the next century.

Fusion, however, is still in the future, and it must be adapted for spaceflight after it has been achieved experimentally on earth. Although there are exceptions everywhere in science and technology, it generally takes about thirty years from the time a result is achieved experimentally to reach the point at which it is easily and readily applicable. There exists, however, a type of fusion engine that uses the only form of fusion successfully developed to date: the hydrogen bomb. It is referred to as the "nuclear-pulse rocket," or the "pulsed-fusion system."

The basic idea of the pulsed nuclear rocket is incredibly simple. Propulsion is derived from the detonation of low-yield hydrogen bombs behind the vehicle. The plasma cloud and shock wave from the exploding bomb push against a gigantic plate mounted on even more gigantic shock absorbers. The plate smoothes out the timed detonations of the bombs and provides a nearly uniform thrust. The idea sounds undeniably crude, but it is entirely practical from a theoretical viewpoint. An automobile may be looked on as being propelled by a long series of explosions, any one of which, were it to happen in the open, would make a serious bang. Even a teaspoon of gasoline, if treated exactly right, can make a healthy explosion. The explosions in the car take place in the combustion chamber, and the resulting force is transferred to rotational motion by the pistons and crankshaft. When the rotational motion is further worked over by the transmission, the result is a smooth, powerful ride, not a series of jarring explosions.

Nuclear-pulse engines have two major things going for them: a hydrogen explosion is to date the only perfected way to convert the energy from a fusion reaction on a large scale, and development costs of the "fuel" would be minimal. Bomb technology has been successfully developed for decades for a much less civilized purpose. This is not to imply that a nuclear-pulse-engine starship would be easy to design or construct, but it is correct to say that it is within what is called "present technology," whereas laser-fusion engines are not.

The idea of a hydrogen-bomb-propelled vehicle was first described by Stanislaw Ulam and Cornelius Everett at Los Alamos in 1955. The concept was later picked up by the General Atomic Division of General Dynamics Corporation, and theoretical work was begun in 1958. Not longer afterward, NASA was established and was assigned responsibility for all nonmilitary space projects. Unfortunately for Project Orion, as it came to be called, NASA was under the influence of the chemical-propulsion groups and was determined to avenge the humiliation of *Sputnik* as soon as possible. Chemical fuels, it was felt, were the way to do that, and the idea of driving spaceships by detonating nuclear bombs seemed much too impractical. Antinuclear political forces were also allied against Orion.

The project was caught among several Catch-22 regulations. It was technically under the Defense Department, but NASA was supposed to have responsibility for nonmilitary programs. The Air Force was willing to help, but by law it could do nothing unless the project were military. Since NASA didn't want Orion, and the Air Force was having to justify the expenditures for Orion as "military," the situation was impossible. Project Orion struggled on from 1958 to 1965, but it never saw development or testing of anything but conventional-explosion models. The immediate applicability of the Orion concept—the nuclear-pulse engine—was to the planning of 100-ton-payload vehicles which could make a fast trip to Mars. The vehicles were small enough to be lifted by Saturn rockets, and the price was low enough that almost half the cost was in the Saturn boosters. An Orion-powered ship could reach Jupiter in two years, Saturn in three. Some of the planners at Orion thought the concept was so practical that mankind would be on Mars in 1964, on the moons of Saturn by 1970; they were overly optimistic. Still, the most intriguing possibility was the application of the Orion nuclear-pulse system to a starship.

Freeman Dyson, who had been with the project almost from the first, developed several starship designs using the nuclear-detonation/pusher-plate concept. A 1968 design of Dyson's involved thousands of hydrogen bombs exploding beneath the ship every few seconds in flight. He calculated the ship could reach 3 percent of light-speed. With a payload of 45,000 tons, Dyson's starship would have reached Alpha Centauri in about 130 years.

A variation of the nuclear-pulse system has been designed into the British Interplanetary Society's starprobe

Daedalus. The engine, an extrapolation of present technology, would push the starprobe to Barnard's Star in 47 years. The speed would be around 14 psol, and the unmanned payload would be 500 tons. The fuel would be 30,000 tons of helium-3 and 20,000 tons of deuterium.

Both Freeman Dyson's starship and the *Daedalus* probe were designed for one-way trips. Both would arrive at the destination at very high rates of speed. The Dyson ship would enter Alpha Centauri's system at 5500 miles *per second. Daedalus* would cross Barnard's system in less than 60 hours after arrival. If either of the two designs was changed to include deceleration at the target star, the trip would be much longer: 100 years for *Daedalus,* 250 years or more for the Dyson ship.

The first proposal for a starship that could return to earth with a *round-trip* time of less than 100 years was made in 1964 by Dr. Robert D. Enzmann of the Raytheon Corporation. Dr. Enzmann presented his proposal to the New York Academy of Sciences. It subsequently became known as the "flying iceberg."

The Enzmann starship was 2000 feet from stem to stern. The fuel was a frozen ball of deuterium a thousand feet in diameter, weighing 12 million tons. It was this frozen ball of deuterium that won the ship its unusual nickname.

The main body of the starship was 1000 feet long and 300 feet in diameter. At launch, the crew would number 200, with provision for 1500 to 2000 people expected to be aboard the ship when it reached its destination. Enzmann believed his starship could reach Alpha Centauri in 50 years, although other scientists argued it would take closer to 100 years. If a sufficient quantity of deuterium could somehow be found at the destination, the ship could return to earth. Fuel could be found in comets, or in any other source of ice or water, since the cosmic abundance of deuterium is 1 part in 10^5 hydrogen.

The Enzmann ship used the nuclear-pulse concept.

It had eight "Orion-type" engines on the aft end of the long main body. The fuel would be delivered to the engines through a central core of the ship from the forward frozen fuel cell. Surrounding the central core was a complex design of 700 smaller compartments broken into living quarters, recreation areas, workrooms, and storage bins. The ship could be spun on its axis to provide artificial gravity for the crew. It was to be equipped with small chemical rockets for exploration of the destination star system.

Dr. Enzmann had originally suggested a fleet of the starships be sent out in the interest of safety and of a higher probability for completing the mission. If there were no way to return, the ships could, it was hoped, find a habitable planet on which to begin a branch of the human race. The Enzmann ship could theoretically reach 10 psol and perhaps greater speeds if the efficiency could somehow be improved.

What the enterprise of Freeman Dyson, Robert Enzmann, and members of the Project Daedalus team of the British Interplanetary Society have shown is that in one form or another, fusion ships are the best chance at the stars which can be proposed today with a reasonable assurance that the technology will be available in the foreseeable future. There are other possibilities for propulsion, some less desirable than fusion, some more desirable, but all are so far in the future that we cannot predict with any certainty when they might become available. Among the other possibilities are solar-sail; laser beams propelling a ship from space-based arrays; the interstellar ramjet; antimatter (or the M-AM drive, for "matter-antimatter"); and drives which might use the fabric of the universe for propulsion.

Going to the stars by using both a laser beam and a solar sail is a fairly recent idea. The sail ship is basically the same as that discussed for years, "the solar sail," but the impulse imparted to it for acceleration is a laser beam instead of solar-light radiation. Laser-beam star travel was first proposed in 1966. The idea was improved on by a Canadian engineer named Phillip Norem in 1969. He proposed a sail 150 miles across, made of very thin plastic, such as Mylar. The spacecraft, a 1000-ton probe, would travel 20 miles behind the sail attached to shroud lines to permit adjustments.

In Norem's plan, a group of lasers would be based in the asteroids; the lasers would be aimed at the sail. The push of the laser beam, in addition to the solar-radiation pressure up to a point, would allow the craft to accelerate to nearly 30 psol in about 10 years. At the 10-year stage of the voyage, the ship would use the background magnetic field of the galaxy to perform a 180-degree turn, which would take 20 years, and would approach Alpha Centauri from behind.

As the ship approached the Centauri system, a second group of asteroid-based lasers 25 times more powerful than the original group would be turned on and aimed at the sail for 10 years. This would decelerate the vehicle to a safe, reasonably slow speed for entry into the star system. R. L. Forward of the British Interplanetary Society also examined the laser-sail possibilities and proposed that the space-based laser array be about 200 miles across so that the laser beam would not diverge too far at distances up to 2 parsecs.

In addition to some of the more obvious difficulties with the laser-sail proposal (where to get the powerful lasers and how to supply them with energy; would a sail 150 miles across tear from fragment collision or disintegrate from interstellar bombardment of tiny particles? would a 180-degree turn utilizing the background magnetic field of the galaxy work?), the trip time to Alpha Centauri, one way, would be at least 40 years.

The American physicist Robert W. Bussard proposed yet another variation on the starship which captured the imagination and has since seen a half-dozen incarnations in a score of articles. Like the laser sail, its primary attraction was that it avoided the harsh realities of carrying along a gigantic fuel load. It was another "free lunch" idea and it was called the "interstellar ramjet."

The principle of the ramjet engine was first explained in 1913 by a Frenchman, René Lorin.

A ramjet engine is essentially a combustion chamber with a forward-facing duct. On earth, a ramjet scoops air into the duct, and the air goes to the combustion chamber, where it is mixed with fuel and burned. The beauty of the idea is that no oxidizer is carried aboard the vehicle; it is free in the surrounding air. The ram effect also increases with speed, so the efficiency of the engine is increased with increasing speed.

In Bussard's ramjet spaceship, the vehicle would scoop up the random hydrogen which exists in interstellar space and feed it to a nuclear-fusion reactor. The reactor would heat the hydrogen and send it out the aft end as propulsion force. The fuel supply would be unlimited. With a 1100-ton spaceship composed mostly of payload (another plus), the size of the collecting scoop required to sweep up the hydrogen would be 1240 miles across.

Bussard envisioned his ramjet scoop as being a magnetic field generated by the starship. As the speed increased, the scoop would be drawn smaller because the ship's efficiency would be vastly increased at higher velocities. At a destination star system, the ship could use the intake field to act as a gigantic magnetic parachute to slow down.

A variation on Bussard's ramjet was proposed by Alan Bond in 1974. In addition to being one of the founding fathers of Project Daedalus, he worked out a hybrid ramjet concept which he called the Ram-Augmented Interstellar Rocket, usually referred to as a RAIR. The ship would be powered by a "conventional" fusion engine using hydrogen isotopes of deuterium and tritium. The magnetic scoop of the ram would be powered by the fusion engine, and it would deliver interstellar matter into the engine to boost the performance. The ship would thus accelerate using conventional fusion means to the speed at which the ram effect would give a significant boost.

The magic of "free" propulsion was so great that all during the 1960s and 1970s, articles and books leaned toward the ramjet interstellar spaceship as being the one way we could successfully achieve about 0.15*c* in a practical way. It was understood there were immense problems in the way of a working version of the ship—the use of extremely high strength materials, producing and controlling intense magnetic and electrical fields over a vast distance, controlling the stability of plasmas in the fields, constructing conventional and workable fusion reactors—but no one asked pointedly whether it would, in fact, work at all.

In 1964, E. J. Öpik, writing in the *Irish Astronomical Journal,* showed that a 1000-ton ramjet starship would require a scoop a million kilometers in diameter. In addition, he showed that the energy losses involved in the ship would be greater than the net addition to the velocity from the ram effect. He concluded that no truly workable ramjet could be devised. Dr. Thomas Heppenheimer followed him in 1978 with a study which also showed the interstellar ramjet would not work ("considering aspects of elementary topics in radiative gasdynamics"). Heppenheimer pointed out that a scoop the size of half a light year would be needed if the plasma in the combustion chamber of the ramjet were "optically thick." If the plasma were "optically thin," then the radiation losses would exceed the energy gain. He concluded, as did Öpik, that the interstellar ramjet was not feasible.

Beyond the perfection of fusion for interstellar flight lie the few real alternative possibilities for rapid travel among the stars. The most talked about of these is the matter-antimatter starship—the M-AM. What is generally called a "mass-annihilation rocket" is one that converts all of its fuel mass into energy: all nuclear rockets employ a form of mass-annihilation but they are not 100 percent efficient in the conversion, via $E = mc^2$, of the fuel to energy.

If a system could be devised that would convert 1 pound of fuel entirely into an exhaust beam, the result would be 5 billion times the energy released per unit of mass in the best chemical rocket. The "100-percent conversion" is the holy grail of the rocket-technology groups (and about as hard to come by as the original, some say).

Total mass annihilation is theoretically possible, and it has been observed to happen in particle physics. Antimatter was first predicted by the English physicist Paul Dirac, and the discovery of the positron in 1932 verified the existence of a particle-antiparticle symmetry in nature. Antiparticles were produced during advances in linear-accelerator technology and were named antiprotons, antideutrons, and antialphas. It is now known that there are antiparticles for all known particles. The antineutron is almost identical to the neutron except that it has the magnetic field reversed. An antiproton is hardly distinguishable from a proton, but the charge is negative, just as the antielectron has a positive charge instead of a negative one.

The particles of normal matter are held together by what has been described as "quantum mechancial glue." Antiparticles are held together by, logically enough, "antiglue." Each glue is a solvent for the other, and when a particle of matter and a particle of antimatter come together, the glues dissolve and the energy of the two particles is released. The mass of both particles is converted 100 percent to energy.*

*Ion-drive, which with current technology may achieve Isp 3500 sec, is also not a desirable starship engine, although for manned missions to Mars and the like, it is the current system of choice within the solar system.

A typical annihilation occurs when the proton and antiproton are brought together. The first result is positive, negative, and neutral pi-mesons. The pi-mesons aren't stable, and they decay into such interesting things as muons, neutrinos, and gamma rays. The muons decay into positrons, electrons, and neutrinos. The final product of the reaction is "normal" matter in the form of the electrons and positrons and a great deal of gamma-ray radiation. The remains of the reaction have about half the total annihilation energy, the rest being lost in neutrino form. The resulting energy is far greater than that in a nuclear-fusion reaction.

The problems in the M-AM starship only begin with producing antimatter, a hard-enough task. Part of the energy release is in the form of neutrinos, which are superparticles that can penetrate anything and everything and refuse to be directed by electric, magnetic, or any other sort of fields we now imagine. Focusing a gamma-ray exhaust beam is also not an easily realized task.

Antimatter is now being produced in very tiny quantities. Proton beams circling in giant accelerators dump protons with high energy levels into tungsten targets. As the protons strike the heavy tungsten nuclei, their energy is converted into a jumble of elementary particles including antiprotons, antineutrons, antielectrons, and gamma rays, all moving at the speed of light.

Magnetic separators route the antiprotons into a large, doughnut-shaped storage ring. In 1979, at Centre Européen pour la Recherche Nucléaire (CERN), scientists managed to store 200 or 300 antiprotons for several days. Planned experiments in the United States at Fermilab are aimed at slowing the antiprotons down from near the speed of light so that containment tests can be carried out.

If the antiprotons and the antielectrons created at the tungsten targets were slowed down and put together, the result would be *antihydrogen* (since an atom of hydrogen has a single proton and a single electron, antihydrogen has, not suprisingly, one antiproton and one antielectron). Feeding ordinary hydrogen fuel to antihydrogen in a starship engine would theoretically produce a 100-percent conversion to energy and a fantastic performance with a relatively small fuel load.

Some studies of antimatter power for starships have concluded that the best "mix" would not be 50-50. At Jet Propulsion Laboratory in California, a technical memorandum on future propulsion systems proposed that the antimatter mass used as the energy source would be negligible when compared with the mass of the propellant. Figures indicating about 4 tons of "reaction mass" for every ton of payload would work, with the reaction mass heated by antimatter annihilation. The JPL studies listed water as a "reaction mass," but hydrogen could be used, as could nearly anything else: rocks, garbage, tin cans, dirty socks, and junk mail. The amount of antimatter needed to heat the reaction mass varies with the payload and velocity requirements. For journeys within the bounds of the solar system, the antimatter used would be surprisingly small. A complex 1-ton unmanned probe sent to Mars would consume 4 tons of water and about 1 gram of antimatter.

The technology of antimatter production is within our grasp now, as indicated by the results at CERN mentioned earlier. But there is a vast distance between what is now the world's supply of .000000001 gram antimatter (and that cannot at present be stored for any practical length of time) and the tons that might be necessary to power a starship. Though the movement of high-energy technology into space will expand the possibility of antimatter production in the future, there are few good guesses when antimatter might be available for use in a starship. Present storage methods for antimatter use magnetic rings, but it is unclear if that is the best way.

Other theorists and investigators have proposed gravitational fields and electrical nets.

The hope for rapid development of antimatter as a possible starship-propulsion system may lie in what have been called "beam-particle weapons." The principle of the beam-particle weapon is this: it sends out very short, intense bursts of high-energy protons which can burn up anything in the beam's path, such as unfriendly satellites, ballistic missiles, space shuttles, and the like. The groundwork on these weapons is well advanced, and it has been claimed that the Russians already have one that is operational or nearly operational. The basic idea has been around ever since Emilio Segrè used the Bevatron to make antiprotons.

What is important in the development of beam-particle technology is the high energy involved. A beam-particle weapon of sufficient size could throw a beam of protons with levels 10 to 15 times the level necessary to make antimatter. But the problems with BPWs and the production of antimatter are huge. Production of 1 kilogram of antimatter, assuming perfect efficiency, which never happens, calls for 50 billion kilowatt hours of energy. This is the approximate equivalent of a very large power plant working at full capacity for 5 years. If all of the energy used in the United States annually were to be turned for one year to the production of antimatter, the product would be about 300 pounds (150 kg).

The only known source for such gigantic energy requirements is the sun. Its output could be utilized by great solar collectors at some point to manufacture antimatter. Whether that will ever come about is uncertain, even if some way were found to store the antimatter thus produced safely. That giant solar-energy collectors—"sunsats"—will be put in space eventually is generally accepted today. Although they are technically feasible, they would cost about $85 billion for a 5-gigawatt system even when they are in full production. (The $85 billion does not take into account what inflation might do to costs until the time when sunsats or powersats might be built.) How long it will be until solar collectors might be diverted from energy production for earth to investigation and production of antimatter is highly speculative.

Another suggested form of propulsion is the photon rocket. A photon rocket is a form of mass-annihilation vehicle, as is the M-AM. As usually conceived the exhaust beam, composed of photons, would reach Isp = 3×10^7 sec, which is the speed of light. The exhaust beam would be radiation. The most common form of photon rocket is a flashlight. If held out in space, it would drive the holder to the nearest star (always provided it were pointed in the right direction; in this case, opposite to the direction of Alpha Centauri) in several billion billion years.

The game is to build a big flashlight, and it would need fantastic "batteries." The power requirements would be astounding: 668 megawatts per pound of thrust. Now, the Saturn rocket had 7.5 *million* pounds of thrust for a short ride to the moon, so the photon rocket seems far in the future. Even if such fantastic power requirements could someday be met, there would be the problem of how to radiate all that power in the proper direction and how to protect some future crew from the twin effects of radiation and heat. No one has the slightest idea of how to go about designing or building anything like a photon rocket, although it was once remarked (with tongue in cheek) that it must exist, since it was entirely possible that quasars—which are point sources of intense radiation exhibiting very strong red shifts, indicating high velocity—were the exhausts of photon rockets.

Projections on the attainable performance of M-AM propulsion systems, including the photon rocket, have been varied because of the many factors involved. A ship with a 10:1 M-AM mixture would yield a high-thrust exhaust mostly of ordinary matter with Isp = 10^7 sec, a performance about 10 times better than that of the best predicted fusion model. The photon rocket has a theoretical velocity limit of the speed of light, Isp $= 3 \times 10^7$ sec, but "in practice," even on paper, most estimates seem to be about 60 psol for the best design.

One final try at the "free lunch" idea of interstellar flight suggests that someday a ship might be able to "push against" nearby stellar bodies. The starship would carry a power source that would generate a field which might transfer momentum from a large local body, such as a star, to the ship. Almost all of the energy would go into accelerating the starship to extremely high velocities approaching the speed of light. It might even be possible to push against individual galaxies or the very "fabric" of the universe and achieve fantastic velocities. Somewhere in the prospect there might be a loophole in Einstein's visions of relativity and a starship might travel at speeds greater than light. But such a prospect, though interesting, is beyond the scope of known possibility now.

NO WAY TO BEAT THE DEALER?

"It is not enough that a thing be possible for it to be believed."

VOLTAIRE

No matter what sweet combination we propose—and some of them are far in our future—we are still limited by the speed of light; indeed, we are limited to fractions of the speed of light. Is there no way to beat the dealer? By what ingenuity could we challenge both the seeming solidarity of Einstein's Relativity theories and the realities of distance? Is nothing sacred?

The possibilities of magic and mind travel, or "teleportation" may be far-fetched, but the long road to the stars may not be as long as it seems. If a fifteenth-century man were handed a bowling ball and asked to describe it, he would say that it was a solid sphere, apparently impenetrable. What he might make of the finger holes might be interesting but not germane to his idea that it was a very solid object. If he were then told that, though it appeared solid, it was made up mostly of empty space, he would reply, "Nonsense." If that were true, he might point out, he should be able to push a finger through it easily. He could not do so, and therefore the theory that it was composed of empty space was silly.

A bowling ball *is* composed mainly of empty space: the gaps between the atoms. Modern atomic theory has made a shambles of the ancient ideas of form and matter. But who is to say that our current thoughts are the final and absolute answers? They do prove out experimentally so far, but so did the concepts of Newton until Einstein came along and opened our minds. And what is proof for a theory? Physical evidence? Logic? Manipulation of mathematical equations? Simple belief? There is a point at which physics and mathematics almost become mythology, and we might as well talk knowingly of griffons, unicorns, the basilisk, or the Irish luċorpán.*

Mathematics can lead to dangerous fantasy. For example, a scientist once "proved" mathematically that no rocket would ever succeed in putting an earth satellite into orbit. This, at the time, might have seemed "self-evident," but there is much about mathematics which should be self-evident and often isn't, especially when reality creeps in softly and behaves in a contrary way. Thus, since science uses mathematics so often to prove or disprove a possibility, it is a good idea, in a Buddhist sort of way, to question everything and take nothing at face value. This is what abstract mathematicians do, and some

*The word in Irish for Leprechaun; Logheryman in Ulster.

of their abstractions are not simply equations hung in midair. The square root of -1, a thoroughly abstract concept, is the basis from which such practical things as the electric motor and alternating current have been developed.

The necessity of mathematical modeling can be illustrated by Zeno's Paradox. Zeno lived in the fourth century B.C. and was one of the first men to worry much about infinity—in this case, the infinitely small. He produced the argument which showed that arrows can't fly. Zeno began with the idea that at any point in time, an arrow does not move at all. (This seems confirmed by taking ultrahigh-speed photographs of an arrow in flight.) He reasoned that if an arrow did not move in a zero interval, then no matter how many zeros were added together, the result was . . . zero. An arrow, therefore, cannot fly but must always remain fixed in space. Zeno was not in the Greek army, or he would have known better. In fact, his mathematical model was imperfect. The flight of an arrow resolves into a consideration of the convergence of infinite series which the mathematics of Zeno's day was not equipped to handle.

Zeno's Paradox was not completely resolved until Newton invented a new set of fancy squiggles called calculus. The flight of an arrow becomes possible only if an instant in time is not plain old zero but something else. The something other than zero is infinitesimal. An arrow does not move at all in a *zero* interval, but it does move in an *infinitesimal* interval.

This concept of something infinitesimally small, but not plain zero, is important in trying to understand the information handed to us by reality—or by what we think of as reality. For example, there is a class of light atomic particles called "leptons." They demonstrably exist. But quantum electrodynamics theory says that if they do, then they must be dimensionless. They have no structure or dimensional extent even at the highest energies.

This is why leptons are called "point particles." An interesting notion: how does something occupying no space have an existence in reality as we know it? Have we abandoned geometry and common sense both? To keep from arousing the ghost of Zeno, we say that leptons don't exactly have a zero dimension; they have instead an infinitesimally small one.

For every particle there is an antiparticle. We think there is antigravity. We believe in the idea of a minus one and forms of it. Is there an opposite to infinitesimally small? Numbers infinitesimally close, but not quite *infinity?* Minus one is smaller than zero. Are there numbers the equivalent of infinity plus one?

The answer to this pretty headache is, in part, yes. In 1915 Georg Cantor of the University of Halle in Germany published the Theory of Transinfinite Numbers. Even the mathematicians laughed. Then others enlarged on Cantor's idea and decided it should be possible in theory to travel between any two points without going the distance between them. Ignoring the requirements of power, it showed that *distance,* under some circumstances, ceased to exist. This was a strange thought, but so is a lepton.

There are other fantasies dealing with points, distances, singularities, and the like which make the slit-toed satyr more believable. How any of it could be used to maneuver a starship is unclear, but mathematical doodling has made breakthroughs more than once in the history of science.

Changing distance into something else goes back to the hoary science-fiction idea of "hyperspace," that ubiquitous place through which spaceships traveled to avoid the wagging finger of Einstein. Ships vanished from existing space, went through hyperspace in hypertime, and came out the other side right where they wanted, maybe light-years or light-centuries away. But hyperspace is not simply a construct of fiction. It might exist.

Nonspace can be presumed to exist where the real stuff does not. If the universe is curved, then the quickest way from one point to another is not around the curve, it is through the nonspace between curves. Calcutta is closer to New York if travel can proceed in a straight line through the earth instead of over the curved surface.

Einstein believed that mass produces curved space. This has been proved experimentally without difficulty. A large mass curves space more than a small mass. A star produces a degree of curvature, a cluster of stars produces more, a group of galaxies more still, until we understand that the sum total of all the mass of all the objects, including particles, in the universe must by definition produce the shape of the universe. What you see is what you get, which suggests there must be nonspace "between" the curves or on the other side of the curves.

It has also been suggested that normal space-time is riddled like Swiss cheese with nonspace tunnels. If access to one of them could be gained, a starship might cross vast distances of real space in near zero (infinitesimally small) time. It might be true that the speed-of-light limit does not exist in nonspace, that it is also nontime; and there may be other properties at which we have only just begun to guess.

Perhaps our universe is finite, and just outside it there might be another one that has as "zero" our mathematics of infinity. Outside that might be another, and then another. This is not completely beyond the realm of possibility. The late Abraham Robinson of Yale University proposed that mathematics could be thought of as similar to changing magnification on a microscope. An infinitesimal point on a scale called a "monad" could be magnified into another scale resembling the original. At the far end of its range, it would have a new sort of infinitesimal number.

Wandering around in this mathematical quicksand is interesting, but no one has come up with any sane proposition to explain how a starship could cross nonspace or other universes by the flick of a switch or the push of a button. The only way we think it could be done involves more travel than going to stars nearby with more conventional means.

That route is through a black hole, a region of space where all the known laws of physics, and some of Einstein's, seem to break down; a place where distance is eliminated and space and time have inconstant or changed meanings.

Of all the constructs of the human thought process, nothing in the imagination is as bizarre as the black hole. It is a gap in the fabric of the universe where, having entered, nothing can escape the way it came. It is a maelstrom with a gravitational field so strong that even light is caught within it. It is such a massive, infinite thing that the space around it is warped and torn into a new shape, and time is distorted. There may be a hundred black holes in our galaxy. There may be a thousand, or there may be a million. We don't know.

In 1798 the French astronomer and mathematician Pierre Simon de Laplace made a strange prediction. He said an object sufficiently massive and dense would become invisible because its escape velocity would exceed the speed of light.

A tiny asteroid a few miles in overall size and of normal rock composition possesses an escape velocity so low that a man walking has enough speed to jump out of the gravitational field involved. An astronaut on one of the small Martian moons could run fast enough to reach escape velocity and become a satellite. A larger object, such as a planet, has a much higher escape velocity. That of the earth is 25,000 mph; of Jupiter, 135,000 mph.

A star would have a correspondingly great escape velocity and a dense star a greater one. A sufficiently massive body would have a gravitational field so strong that nothing could escape it even at the speed of light—including light itself.

The question of how such an object could be formed was not taken up until some years after Laplace. A black-hole is the final act in the death of a star. Even now it has been claimed, somewhat tongue in cheek, that "there is more hard evidence for life after death than for the existence of a black hole."

The intriguing part of the business of black holes is that the warped space-time near the singularity may open into another universe. In 1935, Einstein and a colleague, Nathan Rosen, published a paper predicting the existence of a corridor between this universe and another. This was the Einstein-Rosen Bridge. If something came into a black hole in a certain way, it could go through and enter a different universe parallel to our own. In the other universe would be another black hole with a singularity. Between the two singularities the space bridge (sometimes called a "worm hole") would permit travel between the two universes which would otherwise be inaccessible to each other.

The relevance of all this to starflight, or at least to the possibility of it, is obvious. If a starship could use an Einstein-Rosen Bridge successfully, it could travel to a distant point of the universe in an instant or at least in an infinitesimally small interval. It might become a time machine if some way could be found for it to return to the original point of entry in our universe—something which current theory says can't be done.

There could be a future time in which black holes were mapped as gateways, and starships would simply head for the nearest gateway which opened close to the desired destination. But for all of this to be possible, a starship must first enter a black hole and be able to survive the trip. Neither Einstein nor Rosen thought such a bridge could exist *in reality,* as opposed to mathematically.

Black holes introduce strong tidal forces that would tear apart any material particle, be it a spaceship or an atomic nucleus. In addition, N. D. Birrell and P. C. W.

Davis, of King's College, University of London, believe there are effects on the interior of a black hole from certain quantum-force fields. When these fields are taken into account, the Einstein-Rosen Bridge is "destroyed." Though the mathematical reasoning of Einstein and Rosen applies specifically to nonrotating black holes, there is nothing in it to suggest that *rotating* black holes are exempt from the force fields.

What a starship crew might see from their bridge windows as they entered the black hole is open to serious conjecture. (So is the idea that they could be there in the first place.) One idea is that they might, indeed, see another universe. Another, much stronger idea is that if they were not totally destroyed, they would see a highly distorted view of the universe from which they came.

Current astrophysics suggests rather strongly that travel through black holes is impossible and that Einstein-Rosen Bridges do not exist. There is, however, the "naked singularity." A naked singularity is a space-time singularity that is not hidden inside an event horizon as is the "ordinary" black hole. It is visible from infinity. If the gravitational collapse of a star produces a naked singularity instead of an event horizon (or in addition to an event horizon), what has been theorized about black holes may not apply. No one has any idea what the nature of this sort of gravitational collapse entails. Perhaps only in this one instance could a singularity be used for starship travel. Or, at least, it is difficult to say absolutely that it cannot be done.*

Considering whether a starship will ever try to enter a black hole brings us back to the problem of trip time. The only object that has been tentatively identified as a black hole is orbiting a star in Cygnus, innocuously named HDE 226868, at a distance of 6000 light years.

*Some scientists have theorized that there is a "cosmic censorship process" in which naked singularities cannot appear or cannot occur in general relativity. The hypothesis is unproved, but there is some reason to believe it may be correct.

We have no way of getting there at faster than light-speed, and the best we could do with any of the propulsion systems so far described would be a one-way trip viewed from earth of 11,000 years to reach the entrance to the black hole.

Even if such a trip were eventually undertaken, and it turned out that a bridge to another universe did exist, there is no reason to suppose that a planet would be waiting on the other side of the bridge in another universe. The exit point might be thousands of light years from any nearby star system.

There have even been suggestions that black holes could be constructed in nearby space at a convenient distance—about 1 light year from the sun. It is at least theoretically possible to do so, but how such a feat could be accomplished is beyond any technology we can envision. The idea would be to construct a black hole near the sun, then send equipment through the black hole and construct another one on the other side. We could then have two-way traffic.

If rearranging the distances to the stars is not easy to do, there may be ways in which the barrier of the speed of light can be broken. Could mankind travel the universe in time measured in days, weeks, and months rather than centuries and millennia? The answer, as it is with the question of distances, is a tentative "possibly."

The first clue to the approach is in the Special Theory. It is often assumed, incorrectly, that the Special Theory states that the speed of light cannot be exceeded. What it really says is that the speed of light cannot be exceeded, or even matched, *by an object with mass.* It has a great deal to say about objects *without mass,* and it predicts they always travel at the speed of light, never faster or slower.

Discussion of the speed-of-light barrier to this point has been concerned with objects that have mass: subatomic particles, atoms, molecules, and starships. But there are particles with what is called "zero rest mass" relative to the universe. A photon is one such. It always travels at the speed of light and so defines it. Photons can be considered elementary particles for some purposes, including this one.

When a photon is produced from a slower-moving particle that has mass, the photon flashes off in an instant or in an infinitesimal interval at the speed of light with no discernable interval of acceleration. Similarly, when a photon is absorbed by a particle with mass, the speed disappears instantly without an appreciable time for deceleration.

The conclusion, of course, is that somehow a starship with all of its mass might be converted to photons. The ship would depart without hesitation at exactly the speed of light. At the destination the photons would be converted back to their original form and the passengers and crew would go their gentle way.

There are some problems with this idea besides the basic one of how it could be accomplished. During some part of the conversion to photons, a force would have to be introduced that would keep the photons together as a group; otherwise they would depart in all directions, and there would go the starship forever. Something called the "uncertainty principle" seems to forbid the accurate reassembly of the atoms of the starship and crew. At best, the end product might be a ship full of corpses.

There is also the problem of the principle of baryon conservation, a little-known law of physics. The law implies that particles with mass cannot be transformed into photons.* However, the principle of baryon conservation has been questioned under some parts of the General Unification Theory, so it might be possible to produce the photo-conversion ship at some future time. Unfortunately, the trip could happen only at the speed of light, if

*This law is obeyed in M-AM annihilation because particles and antiparticles have baryon numbers of opposite sign.

at all. It has also been suspected that the amount of energy required to convert a starship to photons might be about the same required to send the starship to the speed of light in the conventional way, and there would be no advantage at all.

One of the older theories of gravity proposes a theoretical particle called graviton. One component of the theory suggests there is a unit called the "two-spin graviton." If the concept of gravitation is correct, such a graviton would have zero mass.* There may be starship possibilities in the zero-mass graviton, but how to turn it to the advantage of starflight has not been suggested. Antigravity, much loved by science-fiction writers, is also a possibility inherent in the gravitation theory; but there is some theoretical evidence which suggests that antigravity, if it exists, has effects that cannot be explored at the level of current experiment.

None of these ideas—the photon-conversion starship, the graviton drive, or some sort of antigravity drive—is a prospect for the near future. There is, however, the "tachyon." If it exists, it would, by definition, always travel *faster than light.* What about Special Theory? Is there a particle which breaks the laws of Einstein? Not exactly. A tachyon is an object with an "imaginary mass," a mathematical possibility only.† If there were an object with mass expressed by the mathematical concept of an imaginary quantity, then for Einstein's equations to remain valid, such an object would have to travel faster than light.

In 1962 two physicists, O. M. P. Bilaniuk and V. K. Deshpande, along with E. C. Sudershan, showed that the Einstein equations allow for the existence of objects

*The same theory holds that positive gravity is a force of infinite range, but antigravity is short-range by comparison.

†Theory indicates that tachyons may not interact with normal matter and so cannot be detected or made use of. The moral may be that Einstein is not always right, but he is never wrong.

which have the imaginary quantity of mass. Another physicist called the particle a "tachyon," from the Greek word for "swift." It was pointed out that the speed-of-light limit existed for the tachyon, but the ceiling was now the floor. In the ordinary world of particles with ordinary mass ("tardyons," for slow) the upper limit is the speed of light. For tachyons, the *lower* limit is the speed of light. In some ways, the tachyon is a reflection of the idea that all mathematical quantities have an opposite somewhere.

Apart from shortening the distance by space warps and ducking in and out of nonspace at will, or driving into the aperture of a black hole, the tachyon is the only real possibility for exceeding the speed of light and making star travel practical *toward the more distant stars.* It deserves some serious attention.

No one knows if tachyons exist. The idea that they do does not violate Einstein's equations, but as in Voltaire's remark at the beginning of this chapter, is it only necessary that something be possible for it to be believed? There are serious arguments against the existence of tachyons, and, of course, no one has yet detected one.

No one had ever detected neutrinos either; but for more than 25 years they were believed to exist solely on the basis of their mathematical possibility. There is, however, a distinct difference between the neutrino and the tachyon. The possibility of the neutrino was accepted because its theoretical existence explained certain observed events. There is no event that would require the existence of a tachyon to explain it. There is only the tinkering with squiggles on the chalkboard.

A tachyon may travel as fast as 1 billion times the speed of light, according to one physicist, Gerald Feinberg. At the other end of the scale, a tachyon, if slowed down completely to the speed of light, could pass out of existence. Detection, so far, has revolved around the use of a hollow sphere made of lead in which there is a very high vacuum. The high-energy particle bombardment of the few remaining elements in the vacuum is expected to produce a tachyon. The tachyon would travel in the sphere at ultrahigh multiples of the speed of light and produce a telltale blue glow called "Cherenkov Radiation."

Theory indicates that Cherenkov Radiation would only happen in a vacuum if bombarded by strange particles moving faster than light. So far, Cherenkov Radiation has been noticeably absent from experimentation aimed at detecting tachyons.

Tachyons might be produced when cosmic-ray showers hit the atmosphere. A tachyon thus produced would begin its travel at the edge of the atmosphere and arrive on earth before the slower-moving (at speed of light) cosmic ray that caused the event. This characteristic of tachyons—that they would precede the events which created them—is one of the arguments against them. There is the Law of Causality (cause must precede effect in time), and tachyons clearly violate this law. Still, there have been a few experiments involving cosmic-ray showers that indicate some sort of event is happening at the ray detectors *before* the cosmic rays strike. Although the evidence is tenuous at best, it might be proof of tachyon existence.

The obvious application would be a tachyon drive. A star-ship would use conventional propulsion methods to approach the speed of light; then the tachyon drive would take over, and the starship would depart the solar system at a velocity much in excess of light on the way to far-flung stars. But how could such a drive be constructed?

A conventional force, Einstein implies, cannot take a starship to the speed of light. Nothing has changed that, just because tachyons exist. In some unknown fashion there would have to be a bridge between the sub-light drive and the above-light-speed drive. At some point, however, the velocity would be exactly that of light—which is "impossible," according to Special Theory, for objects with mass.

A faster-than-light drive is easier to theorize than to produce. And even if one existed, there are severe problems in the way of its use. Anything traveling at vast multiples of *c* would suffer damage from the smallest particle it struck. The calculated energy of the most humble particle at many times light-speed is enough to make any potential astronaut look around for a slow boat to Centauri.

If one of our future generations solves the secret of a star drive, and at the same time solves the secret of how to avoid particle collision at greater-than-light velocities, there is still the question of navigation. While it is not too difficult at sub-light-speed—the stars can still be used—once *c* is surpassed, stars can no longer be guides.

In considering faster-than-light travel, there is one other peculiar phenomenon worth mentioning. Radar people often talk about "wave guides." These are rectangular aluminum or copper tubes along which high-frequency radio waves can pass. When the waves do so, they create electromagnetic patterns. The patterns, it can be shown, go along the wave guides at speeds greater than light. As patterns, they are not energy; nor are they matter. They could best be described as "configuration." If they were changed, they would travel at less than light-speed. They exist, though they go faster than light, but do not violate Einstein's concepts. The "proof" of this, however, involves highly complicated mathematics.

Although there are many ideas about how to decrease the distance to stars or to go faster than light, Einstein's theory still holds dominance over most of our proposals. And those that at first glance seem to invalidate Einstein upon closer examination do not.

The dragon in all these pleasant palaces of mathematical dreams (nonspace, curved space, black holes, Einstein-Rosen Bridges, photons, and tachyons) is the problem no one has yet solved: how to turn any of them into a practical method for constant commerce between the stars, or even for primitive exploration.

THE METHUSELAH ENZYME & OTHER THOUGHTS

*"Ich lehre euch den Übermenschen. Der Mensch ist etwas, das Übermenschen werden soll."**

NIETZSCHE, *Also Sprach Zarathustra*

The third part of the distance/speed/time equation of starflight deals with the human lifetime and, ultimately, with biology, genetics, and medicine. Assume a given lifespan of 80 years. A crew would require 30 years to be prepared for a voyage, leaving only 50 years for flight and exploration. Senility might set in first. If the speed of

*I teach you the superman. Man is something to be surpassed.

light cannot be reached or exceeded, then one-way trips may be inevitable.

Propulsion devices in the near future will not allow speeds of more than a few percent of *c*. With fusion and antimatter starships, at least sizable fractions of the speed of light are attainable, but to do so requires an unthinkable fuel load—more than we or any near-future generation can produce. To bring a starship design to a reasonable possibility, the fuel load must be smaller and the trip consequently longer. At 10–15 psol, only Alpha Centauri can be reached in the given lifetime of a crew; their children would have to make the return trip.

For distances greater than the Centauri system, the only chance, at 10–15 psol, would be for the descendants of the original crew eventually to enter the solar system of destination, perhaps hundreds of years after the ship left earth. This multigeneration starship is not a new idea. J. D. Bernal, a British physicist, suggested it in 1929. He called the ships "interstellar arks."

Each of the arks would be a closed ecological community able to maintain life cycles for generations. In case the principal destination system did not have a habitable planet, the captains would direct the great ship to the next-nearest system, and succeeding generations would investigate it. Eventually the arks would deliver mankind to a new home around another star, where the human race would develop in a way parallel to, but different from, its course on earth.

The interstellar-arks idea led to innumerable science-fiction stories in the days of the pulp magazines, including the Robert H. Heinlein classic *Orphans of the Sky*, first published in 1941 as a two-part serial in *Astounding Science Fiction*. It appeared under the titles "Universe" and "Common Sense." Heinlein anticipated most of the potential hazards of multigeneration travel in this single, short novel: anarchy in one generation, cannibalism, machinery malfunction, genetic problems, mutations, and mutiny. In his story the people aboard the starship completely lost the reason for their journey, and the earth became a myth; the ship was the whole universe; the crew believed there were no stars toward which they were traveling.

Multigeneration travel is an answer to the low speeds anticipated at the beginning of starflight. The genetic parameters of the human race are well known, the ecology needed is within current technology, and the amount of consumables necessary for a trip can be estimated with some accuracy.

There are several variations on the multigeneration starship, one of which was proposed by Arthur C. Clarke some years ago. Clarke envisioned a starship that would only carry the ingredients needed to produce a human population. Eggs would be kept in carefully controlled artificial conditions until the ship was 20 years from the destination. Robot equipment would then see that the eggs were fertilized from a sperm bank. The resulting children would be cared for by robot equipment and educated by a vast tape library. When the destination system at last came into view, the adult crew would take control and put it into orbit around one of the more promising planets.

The whole system could be too delicate to survive the journey, and the ship might arrive empty. Or radiation could so damage the eggs that the resulting children would be only faintly human. Among the other misadventures such a voyage might produce would be that the ship might arrive early, leaving the first exploration of another stellar system to a group of half-savage children: *Lord of the Flies* in space.

This biology ship does not require that the trip be a one-way excursion, except for the first crew. If they could withstand the psychological pressures of having no parenting other than that done by robots and tapes, they would explore the system and record their observations and those of the instruments aboard the ship. Then the ship would be reprogrammed for a return to earth, an idea thoughtfully introduced by long-dead voices chanting from the learning tapes. If one of the planets in the system were habitable, they might elect to stay there and let an empty ship return. Or they might set course for earth and, when the ship had been completely checked out, commit ritual suicide. Twenty years from the solar system, hundreds of years later, another crew would be artificially born to meet the descendants of the original builders.

This type of starship program is not beyond the reach of current tinkering with the pretty works of evolution. In 1959, Dr. Daniele Petrucci announced that he had achieved test-tube fertilization. His monumental work was quickly condemned by the pope, and the scientist destroyed his results. Meanwhile, in England, the Archbishop of Canterbury was either less concerned than his opposite number in Rome or the scientists were more adamant. They experimented with fertilization *in vitro* (meaning, literally, "in glass"). Drs. R. G. Edwards and Ruth Fowler at Cambridge devised a method of obtaining healthy human eggs by removing them during minor abdominal surgery.

The eggs thus obtained were then fertilized by sperm in culture dishes. The English investigators were able to produce human embryos in very rudimentary stages. The culmination of this form of minor genetic control was the birth of the first test-tube baby in 1979. Earlier, in the 1960s, embryo transplants had been accomplished with cattle and sheep. The success was not, however, a nonhuman birth. The conception occurred in the laboratory, but growth to term and birth took place in a human being. Research is just beginning on the ultimate possibility of a human life conceived, nurtured, and born in an artificial environment. With these developments, the idea of sending out a biology spaceship is not particularly farfetched, although it may be one of the least desirable ways to explore the stars.

Applying genetic engineering to starships does not stop at birth accomplished outside the human body. There is the matter of changing and working with genes.

Prior to 1950, it was known that there were microscopic specks inside our cells—genes—that carried heredity and determined who we are, how we function, and what we look like. A human being run through an analyzer is indistinguishable from an equal amount of bacteria. It is only the *arrangement* of the molecules which distinguishes one from the other. The determination of that arrangement is by means of the hundreds of subunits which compose a gene.

In 1950, it was discovered that the subunits are strung together in a long, threadlike strand of material called "deoxyribonucleic acid"—DNA. DNA is the main substance of the chromosomes, the basic genetic material, the governor of biological mechanisms of reproduction. This new information led Francis Crick and James Watson at Cambridge to isolate something called RNA which carried the message of the DNA in a four-letter "code." They received the Nobel Prize for their model of DNA and the discovery of RNA. It was the $E=mc^2$ of biology.

One of the first thoughts which arose, once DNA became known, was the spectre of a human clone.

There are two types of cells in the human body: body cells and sex cells. The sex cells have only half the required number of chromosomes for reproduction. Reproduction, therefore, takes place when the two sets of sex cells meet. In "normal" reproduction as well as test tube, the two different sex cells, each with half the required chromosomes, combine to start a new creature, whose inherited traits are transmitted by the genetic information of the egg and sperm through their genes.

Body cells contain a full set of chromosomes with the genes basically "turned off." DNA is a biological governor of reproduction, but it can also be thought of as a biological switch. It can activate a body cell into a reproduction cycle. The result would be asexual reproduction: only one parent. It is called a "clone," from the Greek word for twig, which is to imply a whole object taken from a cutting, much in the way some plants can be reproduced.

In 1968, J. B. Gurdon of Oxford managed to produce a cloned frog.* He took an unfertilized egg cell from an African clawed frog and destroyed its nucleus by ultraviolet radiation. The nucleus was replaced by one from an intestinal cell of another frog of the same species. The frog egg was tricked into a reproductive cycle and the result was a tadpole clone. It, too, could reproduce.

In 1979, Karl Illmensee of the University of Geneva reported he had succeeded in transferring the nucleus of a mouse-embryo cell into the egg cell of another mouse. He removed the egg's original genetic material and ended up with a normal—but clone—adult mouse, after using a host foster mother. This was the first successful nucleus transplant into a mammal.

The prime attraction in cloning is what has been called limited cloning. This would involve only specific organs. If the true master switch of cell reproduction is found, then it would be possible to clone a new heart, liver, kidney, etc., and transplant it to the original donor for a perfect match. Human lifetimes could be greatly increased.

A starship crew member is about to lose his heart function through aging or disease. Some of his genetic material is injected into a human egg. The original donor-patient with the diseased heart is placed in a "cold sleep" environment while the egg is put into a Silastic (plastic) uterus. Within the artificial uterus, a computer

*With due respect to Dr. Gurdon, biologists have pointed out that frog eggs are notorious for being easy to work with. A frog egg will grow parthenogenically when pricked with a needle.

monitors the exact hormonal concentrations so that a normal attachment to the uterine wall can be achieved. In our real-life system the chemical balance that enables a human embryo, after a few days, to attach itself to the wall of the womb is very delicate and very hard to accomplish synthetically. But an artificial system governed by computer, with great precision and control might be engineered someday.

As the fetus developed in the artificial environment, genetic repressors would be injected into the synthetic blood that keeps the growing organism going. All development of the fetus would stop, except for that which would be free of the genetic repression: the heart. In a relatively short time a perfect, identical, and practically nonrejectable (since it is genetically identical to the donor's) human heart would be ready for transplanting into the awakened patient. The new organ would be about the size of a child's heart and would have to be helped along with other devices until it grew up.

An alternate scenario might be limited cloning for regeneration. Diseased cells from an organ would be removed and the remaining good cells injected with "on" switches. The organ should regenerate perfectly.

Could clones completely man a great starship? Yes, certainly. A single astronaut-donor could furnish the ship's biology banks with any number of body cells containing a full set of chromosomes. The fertilized eggs, developed in machines simulating human-womb conditions, would grow into clones of the astronaut. The clones could operate the ship for generations until the destination was reached. In this way a direct lineal descendant of the original astronaut would step out on a planet around another star.

This idea has particular merit when it is suggested that a starship could have a crew placed in a frozen state (cryogenic suspension) to sleep out the voyage: the Rip van Winkle effect. The original crew would man the ship until it cleared the immediate area of the solar system. They would then activate the biology section to produce a full set of crew clones. The clones would grow to maturity, man the ship, and protect their "fathers." They could reproduce in the normal way for generations, or they could activate additional clone births after 20 or 30 years and train the new clones. Twenty years out from the destination system, the clone crew would activate the original human crew and bring them back to life for the landing.

With or without clones, the idea of a frozen-sleep journey is attractive. A trip to the nearest star could be made in an "instant"—or almost. The crew would control operations out to an appropriate distance from the solar system; their computers would line the ship up on Alpha Centauri or another chosen destination. They would enter their sleep cocoons and set computer alarms, to be awakened when the outer fringes of the destination system were reached.

After exploring, the crew—hardly any older—would set the ship on a course toward earth, crawl back into the sleep chambers, and return. That they would know no one on earth after a hundred or more years might be the least of their problems. However, it is one way a voyage to the stars could be achieved quickly for the crew and would avoid carrying the millions of tons of air, food, soil, and other essentials that a multigeneration or clone ship would require. The crew wouldn't get bored; there would be no problems with genetics, or with the morality of dooming clones to deadly drudgery for generations to secure their masters a place in the annals of human history. The crew would awake knowing the purpose of their trip and be fresh and ready for exploration.

We are probably closer to cloning than to the perfection of cryogenic suspension. Although medical science has progressed to the point where slowing of metabolism by low temperatures is commonplace for a few hours of surgery, there are as yet no techniques for long-term freezing.

The principal difficulty with freezing a human body is the formation of ice crystals inside the cells. The freezing water expands until the cells are ruptured. There are drugs that will inhibit this process, but the body would have to be saturated with them to allow a huge drop in temperature and the side effects of the drugs in massive doses are unknown. Freezing at several thousand times the pressure of the earth's atmosphere would solve the ice problem, but how would the body survive such pressure?

Cryogenic suspension presents other difficulties that must be overcome before a starship crew heads for the sleep chambers. The radioactive atoms within the human body would decay tissue surrounding them. Since the body would be frozen, no continuing replacement process would take place, as it would if the crew member were alive. No one knows how long a human being could be frozen before the radioactive effects would cause medical complications that would make it inadvisable to revive the person. The time interval might be anything from a hundred years to several thousand.

The equipment aboard the starship would have to be capable of sustaining the cryogenic state without failure unless the crew were gamblers, and assumed a certain failure rate to be a natural part of their lives. Revival from cryogenic suspension is a process about which nothing at present is known.

Studies of those animals that hibernate in nature may eventually result in an "estivation drug" for starship crews. There are some very strange examples of hibernation in nature, including a deathlife state in certain primitive creatures that have been almost completely dehydrated: "cryptobiosis."

Cryptobiosis, for "hidden life," was coined in 1959 by David Keilin of Cambridge University. Organisms that can go into a cryptobiotic state at any time in their life cycle, voluntarily or involuntarily, include a number of bacteria and three groups of invertebrates. The latter are the wheel animalcules that inhabit freshwater ponds; microscopic, wormlike nematodes found in moist soil; and primitive arthropodlike animals called "tardigrades."

It is not yet known how long a cryptobiotic organism can remain in a state of suspended animation. Nematodes have been revived after being kept in a dried condition for nearly 40 years. In one of the strangest studies in cryptobiosis, a piece of museum moss kept dried for 120 years was a home for several kinds of rotifers and tardigrades in suspended animation. When the animals were moistened, a few of them revived. They did not, however, stay revived for long, and died again, this time permanently.

The cryptobiotic process allows for greatly increased life span. Ernst Marcus of the University of São Paulo believes that organisms that might survive for one year under normal conditions can survive for as long as 60 years with alternating active and cryptobiotic intervals. If the process could be applied to humans, the lifetime of a crew without the addition of any other medical advances could be 3000 years.

Another interesting feature of the cryptobiotic state is a resistance to environmental changes and extremes. While in a cryptobiotic state, life forms have survived ranges of temperature which would have been instantly lethal in the active state. In the 1920s, P. G. Rahm at the University of Freiburg, Germany, reported that certain creatures in a cryptobiotic state could survive temperatures far above the boiling point of water and could last for days at temperatures several hundred degrees below zero F. A study at the University of Paris indicated survival at almost absolute zero: .008° Kelvin (about −450°F).

Cryptobiosis has other properties of interest. The amount of X rays needed to kill the life forms within 24 hours is 570,000 roentgens. Humans will die from 500 roentgens in the same period. Even at such a massive dosage, only 50 percent of the creatures died. Some of them can survive two days in a vacuum. They can also live without oxygen (as in a vacuum) and heat, and can lose their body water entirely and still be revived. They do not need water as an absolute essential to life.

Is any of this cryptobiotic process applicable to a starship crew? It is hard to say what might result from biological tinkering. It may be that in some distant future a starship crew will not be "human," as we would understand the term. They might alternately hibernate as cryptobiotic creatures do and be active when needed for running their ship and exploring. With further development in biological engineering, it might be possible to produce a semihuman crew who could last without much oxygen, water, or heat, at near the dead-zero temperature of space and be revived by their computers and hibernation machines at the end of the voyage. The voyages might last 3000 years. Freeze-dried spacemen. Just add water at destination.

The word "bionics" was coined by Major Jack E. Steele, a research psychiatrist and flight surgeon at Wright Patterson Air Force Base in Dayton, Ohio. It is a combination of the Greek "bios," meaning "life," and the suffix "ics," meaning "after the manner of."

The original bionics was an attempt to merge engineering with biology and is not the same as what is now called "biological engineering." The goal of bionics is nothing so piddling as prosthetic limbs and plastic surgery, which were its origins. Bionics seeks to improve nature or replace it when failure occurs.

One of the first bionic devices was the now-famous pacemaker, followed by the artificial heart valve. Not too many years later developments made it possible to contemplate artificial hearts that would be regulated by plutonium-generated pacemakers that would pump blood through plastic veins and arteries. Arms could be powered by transducers and electrodes thinner than a human hair; legs could be supported by artificial knee-caps; knuckles could be replaced with plastic. Not far away was an experimental device that could make deaf ears hear and a brain-implanted device (of electrodes) that could make blind eyes see.*

None of this came easily. The design complications of something like an artificial heart are incredible: the device must be able to flipflop endlessly, surviving the corrosive action of the blood (blood is more corrosive than some acids; it can destroy metals in a few months); it must function in the hellish environment of the heart. Above all, it cannot fail.

By 1977 the most complex of biomedical devices, an artificial heart, had lasted in an animal as long as 186 days; and by 1978 artificial arms were developed that could duplicate almost all of the motions of a human arm and begin to do so as soon as they were fitted.† The ultimate goal was bionic extension of a human being's capabilities.

In addition to spare parts (either cloned or bionic) being obviously desirable for any crew of a starship—in fact, they are necessary—the ideà of bionic extension provides another dimension. The best pilots are, in an almost magic way, welded to their aircraft. It is as if they were a part of the plane, or as if it were an extension of themselves. What if human biology could be wedded to electronics so perfectly that the pilots became one with their vehicles? That would be a real Cyborg, and it would have applications for starflight.

No one knows how long a human brain might live if it

*The ultimate in bionics is a "Cyborg," or cybernetic organism.

†At MIT in 1981, doctors succeeded in growing human skin in a test tube. The material is a replica of living tissue, grown from a sample from the prospective recipient. It will be used in treating burn patients.

were not subjected to the destructive excesses human beings visit on their bodies. Artificially kept, it might live and be perfectly capable of working successfully for more than a century or two. If a Cyborg pilot were implanted into a starship's control and navigation system, he or she could guide the great ship all during the voyage. And it, or another brain, could be assigned to watch over a sleeping crew, monitor the biology repository carrying human eggs and sperm, or even produce clones.

Stretching the lifetime of the human brain is, of course, essential to improving longevity generally. A multigeneration starship or clone ship or hibernation ship would not be necessary if the original crew were able to last the duration of the voyage.

Bionic parts and cloned replacement body parts can extend human lifetimes far beyond what they are now. But what is really needed to stretch the whole process is a "Methuselah enzyme." To find that, we will probably have to discover what causes aging.

There are three general hypotheses that are thought to be plausible explanations for the aging process. All three are based on the properties of the information-bearing molecules in the cell, the DNA and RNA. The first was advanced by Zhores Medvedev of the Medical Research Council in London. It was further developed by Orgel of the Salk Institute. In this view, known as the error idea of aging, the information system represented by the transcription and translation of genetic messages in the DNA and RNA becomes increasingly subject to erroneous information over a period of time. The errors create faulty enzyme molecules which then lead to a rapid decline of the abilities of the cell to function. This is somewhat analogous to an assembly line in which a computer error orders the wrong or undersize parts. The final product might be a complete device, but one with vastly diminished capabilities in performance and service.

This view of aging has been tested in laboratory conditions, but the results have not entirely corroborated it. Most organisms, including humans, have a built-in capacity for correcting errors in the repair process which goes on at the cellular level. Species that live half as long as human beings also have half the repair-capability rate in the DNA.

A second hypothesis, also originated by Medvedev, suggests a genetic basis for aging. It is thought that although the DNA messages are redundant, the built-in redundancy is repressed. If an active gene is damaged, however, one of the reserve redundant genes is put in its place. Eventually, over time, all of the reserve would be used up and errors would creep into the process which would lead to the age changes that we observe in all life forms.

The third hypothesis for aging presumes there is an aging signal or even an aging gene which does its job at the appointed time. It is known that some cells have "death clocks" which operate on a very precise schedule (during growth, for example, cells die and are reabsorbed on what appears to be a fixed schedule), and it may be that aging, as a process, is simply signaled and occurs in the ordinary course of events. This aging gene idea may be a process that happens all through life, operating at different rates in different tissues, and the results of the process are what we call aging and the characteristics of functional decline in various body parts.

Supporting this third concept is the view that aging is an evolutionary, rather than a purely biological, necessity. From the standpoint of evolution, it is necessary for an organism to die after its reproduction cycle is no longer viable so that succeeding generations of the organism can carry out minor evolutionary changes.

A different theory suggests that aging is caused by oxidation. Electrically active chemicals in the body, called free radicals, cause "rusting" of the body tissues. If the oxidation process were halted or slowed, longer life would be possible. There are three antioxidants that might someday be used as part of a Methuselah enzyme: 2-MEA*; vitamin E; and butylated hydroxytoluene, a preservative in common use.

Human lung cells in test tubes have been treated with vitamin E; the result was a doubled life span. Both butylated hydroxytoluene and 2-MEA have been capable of doubling the life span of some cells in some experiments, and mice with diets saturated with the antioxidants have experienced twice the normal life span.

In 1974 researchers in the United States tested a drug that could potentially triple a human life span. It was first tested on flies, and there was some confidence at the time that it might be tested on humans in the 1980s. The intent of this particular line of research was to lower body temperature as a way to slow the aging process. A permanent reduction of the body's temperature to about 86 °F could result in a life expectancy of 200 to 250 years. Unfortunately, the biochemistry of warm-blooded animals has been developed along a vast evolutionary line, and most indications are that a permanent reduction in body temperature would probably result in a less successful biochemisty. A 200-year lifespan as a semivegetable would not be desirable.

By far the most interesting evidence relative to aging appeared in 1980. Leonard Hayflick, working at the Children's Hospital Medical Center of Northern California, examined the process of aging at the cell level. He discovered that when the structural cells of the body's soft tissues were grown in laboratory glassware, they divided over a period of time in the normal fashion and then slowly stopped dividing and eventually died. This suggested that the aging of normal cells in the human body was an innate property of the cells themselves. No matter how the experiment was done, the conclusion seemed the same: the cells had a limit on the number of times they could divide.

It would seem from Hayflick's work that normal cells

*2-MEA is an abbreviation of "2-mercaptoethylamine."

are innately mortal, but there are immortal human cell lines as well. For example, a normal human cell treated with a cancer-causing monkey virus becomes an ageless human cell line. About 600 immortal cell lines are known. The most famous is one called "HeLa," which dates from 1952. It has been in continuous culture since that time. These immortal cell lines are quite different from normal human cells. Their chemical properties are changed, as are the number and shape of their chromosomes. When the immortal cells are injected into laboratory animals, they cause tumors.

The conclusion, according to Hayflick, is the paradox that for animal cells to become ageless (capable of unlimited division), they have to acquire many of the properties of cancer cells.

If the biological clock mechanism of aging is the true explanation, then it may be possible to change the clock's rate. A fruit fly is ancient at the age of 40 days. A mouse is an old man at 3 years; a horse is old at 30. Man sometimes lives to 100, and the tortoise to 150. With a new clock in his body, man might live to 200 or 500 or thousands of years. A starship crew with a lifetime of centuries would make a voyage to the stars a real possibility regardless of the relative inefficiency of its propulsion system. If such an increased life span were combined with cold-sleep processes, stars much farther away could be explored, and even the nearby galaxies might be within reach. A round trip could be contemplated within the lifetime of the crew if the people on earth also had extended lifetimes. The idea is a grand one. Imagine the monumental celebrations of a nearly immortal human race welcoming its brothers and sisters back from the stars after a voyage of 2000 years.

The idea of modifying the human race in very basic ways falls into the realm of what is called rDNA, for recombinant DNA. There is no single approach: rDNA is a group of techniques that can be used for a wide variety of experiments and applications. One of the basic experiments is to take apart a DNA molecule and put the segments back together again—possibly in a different way. The techniques have been responsible for considerable hysteria among the public, and rDNA work has been viewed by some as a Pandora's box. Horrible visions of terrible diseases resulting from rDNA research have been offered by the news media and by critics of the research. To a degree, some of the public apprehension is justified; our history in this century has had a number of examples of the application of science for dubious purposes.

Lassa-fever virus, Marburg virus, and Zaire-hemorrhagic-fever virus are all self-propagating biological agents that have a very high mortality rate in infected people. It is a frightening thought that somewhere, someone is trying to "improve" one of these agents using the techniques of rDNA. On the other hand, rDNA techniques are much like a shelf of chemicals: everything depends on how they are used; limitation of use is the question, not whether they should be used at all. Any self-respecting student in a chemistry class can make a thoroughly dangerous substance if he or she is allowed to.

Modifying something as basic as DNA may bring a solution to the aging process before a starship is ever built. If that is not possible, then other potential avenues are open. Years ago *Esquire* magazine, in one of its more memorable spoofs, ran an article about what future man might look like. The illustration for the article showed gills for breathing; special and additional hands for holding this and that; special growths to cope with pollution, acid rain, and other ills of the supposed future environment.

In fact, some such creature could be created using techniques learned in rDNA research. A starship crew might include a combination of biological products: cyborgs in control and navigation; clones manning other departments; original, "normal" crew members in frozen sleep; banks upon banks of frozen cloned human parts and biomedical-engineering pieces; and a landing party of constructs developed to withstand almost any environment on almost any planet.

So far, the considerations have been limited to physically changing the characteristics of a human being who might be part of a flight crew. But there may be other ways.

The mystery of life and the meaning of reality are all-consuming philosophical concerns in mankind's history. Some have pursued these subjects by means other than cyclotrons, sweet combinations, gigantic rockets, or intricate squiggles on a chalkboard. Paramahansa Yogananda was the first great master of India to live in the West for any considerable period of time. In 1946 he attempted to explain some of the phenonema of the consciousnesss of the perfect yoga as it related to Western thought.

"A perfect yoga," he wrote, "who through perfect meditation has merged his consciousness with the Creator, perceives the cosmical essence as light . . . to him there is no difference between the light rays composing water and the light rays composing land. Free from matter-consciousness, free from the three dimensions of space and the fourth dimension of time, a master transfers his body of light with equal ease over or through the light rays of earth, water, fire, or air."

He wrote that masters who could materialize and dematerialize their bodies and other objects could move at the velocity of light and utilize the "creative light rays" in bringing into instant visibility any physical manifestation. He said this did not conflict with Einstein at all: perfect masters who did things like that also fulfilled the condition that their mass was infinite. Gravitation did not exist for a yoga master in perfect meditation; nor was he subject to the rigidities of space and time.

Yogananda's explanation of light was a much more complex idea than simply a beam of photons that are sometimes (but not always) particles, as Westerners would view it. His concept of relativity is also rather strange, but it crosses two widely different cultures and two languages, and it is clear that he is trying to convey some very difficult ideas with sincerity.

There are many otherwise ordinary people who use yoga to accomplish tasks we would normally think impossible. In the 1960s, a diver descended to 300 feet, in a dive lasting 4 minutes, with no equipment but his lungs. At that depth the water pressure surrounding his body would be measured in *tons.* Divers in suits who have lost their air supply at that depth have been found squeezed into the bronze helmet as a shapeless mass.

Other reports of phenomena resulting from yoga meditation have included extending human lifetimes to 110 years, lowering a human heartbeat to 28 beats per minute, and feats of levitation. Tied in sometimes with yoga have been claims of teleportation—mind travel or astral projection. Yogananda mentioned several such experiences of his own.

Teleporting yourself to the stars by wishing real, real hard you were there is interesting; but it is subjective and, worse, it is nonphysical in the sense that we understand the word. *J'étais là où je pensais être, mais j'étais ailleurs* (I was where I thought I was, but I was somewhere else). What would you do when you got there? What could you build? How would you become what you are now in another place?

It is, however, almost predictable that a starship crew will take advantage of every technique available when voyages to other star systems begin. And the far future may bring a race of humans not only unrecognizable in shape or form but having unforeseen talents of the mind and spirit as well.

In 1966 J. von Neumann wrote a book entitled *Theory*

of Self-Reproducing Automata. In this and in subsequent works, von Neumann proposed that a self-reproducing universal constructor was theoretically possible. Such a machine would be extremely intelligent and capable of making any kind of artifact or product, if it were correctly programmed and if the proper construction materials were available. The most important aspect of the machine would be that it could duplicate itself. A human being can be thought of as a biological von Neumann device with a speciality in performance on earth-type planets.

If such a device were ever perfected, it could be used to extend mankind much farther than the nearest stars. When the von Neumann device/starship arrived at a solar system, it would be programmed to seek out and locate raw materials: debris, small asteroids, and other materials left over from solar-system formation. These materials would be turned into human habitations by the universal-construction machine, perhaps space colonies along lines which have been suggested by Gerard O'Neill, Thomas Heppenheimer, and others. Once a space colony or giant space station was completed, the von Neumann device would synthesize a fertilized egg cell of a human using information in its intelligence. The crew would suddenly be alive and robots, also built by the device, would educate them and help them develop a colony around the target star.

Once the project was in operation, the von Neumann device could duplicate itself from raw materials of the new solar system and send its copy on to another where the human race would also flourish. Since the idea is exclusively a one-way program for colonization of the stars, and the "crew" would not be aware of a trip until it was essentially over, the time required for the voyage to one star or to a dozen is relatively unimportant. It could take a million years to partly colonize the nearest stellar systems, and it might be 10,000 or 20,000 years before any word of the colonization reached earth, depending on the time required to launch a von Neumann device, make the trip to a stellar system, and establish a human station.

A von Neumann device is much easier to talk about than to build. While computer technology is rapidly approaching "intelligence," it is a long way from a self-reproducing universal handyman that can handle the complex operations of raw materials to finished product, retrieve asteroids, build colonies, function as an entire technological civilization from one basic program, and fly to the stars. Even if someday one could be made, there is no guarantee that the synthesis of a living human cell or synthesis of fertilized egg cells is attainable by either machine or mankind. Given our history of tinkering with the affairs of the universe, such a project is not impossible, but when it might happen is beyond our powers of prediction.

PLENTITUDE CANNOT BE RESTRICTED

"Innumerable suns exist; innumerable earths revolve around these—living beings inhabit those worlds."

GIORDANO BRUNO, 1548–1600

Inherent in thinking about travel to another star system is the idea that in one of those distant worlds there might be another home for man. Are there planets that are habitable for us circling one of these "innumerable suns," and what might they be like?

Giordano Bruno did not live to see the invention of the telescope, which would have given him proof of his statement quoted above. He based his conclusions not on scientific reasoning but on the religious idea of Plentitude, which was popular in his century. The Doctrine of Plentitude held that God's handiwork could not be restricted; thus there could be many worlds, many peoples in Creation.

The invention of the telescope produced an explosion of information about the planets which led to a slightly more scientifically oriented theory called the Plurality of Worlds. The planets known so long to the ancients were revealed as solid globes which went around in their orbits of the sun producing "years"; they rotated on their axes, producing "days" and "nights." Some were accompanied by satellites, as the moon accompanied the earth. And, if it were true that the stars were really suns like our own but more numerous than anyone had ever imagined, then it was also possible that each of those suns had its family of planets too.

At one time, all of the planets were believed to be inhabited; this was, in fact, the basis of "plurality." If only one was not, then it followed that each of them would have to be taken as an individual case and plurality was disproved by exception. Astronomers began their studies, inevitably, with the moon. In very early telescopes, the moon had seemed a second earth with seas, mountains, valleys. More improved instruments showed the seas were not seas; there was no water. When the moon passed in front of a star, as it frequently did in its orbit around the earth, the star did not dim slowly as it should if there were a mantle of air around the moon. After a century of observation, not a single change that could be documented by more than one observer had appeared on the lunar surface. Plurality, as a theory, was in serious trouble.

There were those who argued that the moon was a special case and that it had never been "intended" to be inhabited, only placed in the sky to give solace to lonely humanity. The planets, it was maintained, were different and they were universally inhabited.

As telescopes improved further, it became apparent that only Venus and Mars were planets much like the earth. Venus was difficult to investigate because the extremely thick atmosphere obscured the ground. Not until the landing of a Soviet spacecraft in 1975 was anything known about the surface. That left only Mars. It had been known since the invention of the telescope that Mars had polar caps. It also had an atmosphere, and there were dark areas which changed with the changing seasons. Its day was 24 hours, and its year was twice that of ours. The planet received about half as much light and heat at its distance from the sun as earth.

Here was a different case. Mars was active. There were clouds which could be seen in telescopes. There were dust storms and changes which appeared from year to year. The planet was very cold but was probably inhabitable. This was a speculation that would persist in various forms until 1976 when a spacecraft landed on the planet and reported that the conditions were extremely unfavorable to life, and there was none to be found around the landing site.

Practically all speculation about worlds around other stars has been derived from the conditions in our own solar system as we have known them over the centuries. One of the first to speculate about planets around other stars was William Whewell, who wrote a book called *The Plurality of Worlds* in 1854. Some of the ideas expressed in the book were far ahead of his time.

He began his speculation with an analogy, carefully pointing out that his information was fragmentary. The stars, he said, were assumed to be like the sun; and, therefore, like the sun, they had planetary systems. But how much they were like the sun would determine whether there really were planets. One of the few facts about stars that Whewell had at his disposal concerned binary systems, discovered by William Herschel. Binary stars are those which appear double in the sky and are actually linked by gravity. If the period of a binary is known, and the distance is known, then the masses of the stars can be obtained. Binary stars would give a clue to how like the sun the individual stars were. Bessel had measured the distance of the binary 61 Cygni in 1838; and by Whewell's time, some other stellar distances were known, as were the periods of some binaries. The result was the determination of the mass of several stars in relation to the sun.

Whewell concluded that most of the binary systems had individual stars with masses between one and three times that of the sun. He also thought that the colors of binary stars indicated temperatures and natures different from that of our sun. This latter conclusion, coupled with what he felt would be an extremely complex orbital arrangement for planets, ruled out planetary systems around binary stars. What he did not know was what single—nonbinary—stars were like. Were they more massive than the sun? Less massive? The information had vast implications for theories about inhabited or inhabitable worlds among the stars.

Today's discussions of other stellar systems do not differ greatly from those of Whewell's day. We now know which stars are like the sun, but whether any of the stars might have habitable worlds is as much a matter of conjecture and analogy as it was in 1854. But current conjecture and analogy are backed by mountains of statistical probability, and there are figures which seem real. It is wise, however, to remember Whewell's admonition: reasoning from analogy is slippery. All of the current reasoning, no matter how accurate it seems, is based on two assumptions: neither the sun nor the earth is unique in the universe.

Whewell would have understood the modern arguments perfectly. Find a single star which is like the sun, and there lies the best chance for finding a planet like the earth. Determine that life exists elsewhere in our solar system, and you have probably determined that life may exist in all solar systems of a similar kind. The inevitable conclusion from the above cases is that stars are generally like the sun; planets are common around stars; earthlike planets are common in solar systems; and life can exist on most earthlike planets. It was partly to determine whether life does exist elsewhere in our solar system that the Viking spacecraft which landed on Mars in 1976 was equipped for biology experiments.

Stars are classified according to their most abundant elements as determined from their spectra. The classes are related to temperature, which allows them to be compared to each other—and to the sun. Most stars are what are called "main sequence" stars, and they are ordinary, normal, and in the prime of their existence. *They are more or less like the sun.* This information was what Whewell lacked in 1854.

The most exhaustive study on the possibility of planetary systems around other stars and habitable planets was done for the U.S. Air Force in 1962 by the Rand Corporation. It was completed by Stephen H. Dole and called "Habitable Planets for Man." The basis for all subsequent discussions of other planetary systems rests on this statistics-filled but fascinating piece of research.

The Rand study concluded that most slow-rotating stars in the galaxy can have planets; and, therefore, two billion planets exist of which roughly 600 million might be habitable. For stars like the sun, the study reported a chance of 1 in 18 for a planet on which man could exist. The size of stars in this group is 0.89 to 1.04 solar masses. This higher probability for stars that are much like the sun is not surprising since the whole statistical analysis is based on the existence of life on one planet—earth—going around a G^2 star of 1 solar mass. A solar system with two habitable planets was considered improbable, which might explain why Mars is not inhabited and not particularly hospitable to habitation by the citizens of the third planet from the sun.

The estimate of 600 million planets was based on a complex series of factors: planetary mass, age, illumination, orbital eccentricity, mass of the theoretical planet's sun, and so on. If all of these fell within certain parameters, then a planet should be a potential world for mankind.

Since the Rand study was made more than 20 years ago, astronomers and other scientists have come to realize that the question of habitability is much more complex than was at first assumed. For one, the fraction of stars that are double (decreasing the probability that a planet would be formed) is believed to be extremely large.* For another, large planets with substantial atmospheres, like earth, may often suffer from runaway glaciation and runaway greenhouse effects, neither of which was known in 1962.

More recent estimates indicate the probable number of planets in the galaxy on which man could live to be about *880,000,* but guessing the chances for habitable planets involves the specific question of how life on earth arose. No one knows the complete answer, so estimating the chances for life "out there" is accompanied by incalculable potential error. Whether life on earth began by accident is often not as hotly debated as the nature of the accident. How random is the apparently random set of circumstances that caused life to begin forming on earth billions of years ago? If the process were common, it would be possible to calculate the chances for life on a planet similar to the primordial earth; if not, a centuries-old concept is revived—the "anthropic principle"; namely, that we are alone in the universe always and forever.

*A 1976 study by Helmut Abt and Saul Levy at Kitt Peak National Observatory concluded that of 123 stars within 85 light years of earth, 83 had starlike companions; they estimated that many if not most of the remainder had a companion other than a star—black dwarf companions, massive Jupiter planets, etc. The implication was clear that there may not be too many earth-like planets nearby, since large or massive companions in gravitational interaction with a star tend to inhibit their formation.

As long as it is assumed that life began on earth in unique circumstances, it is difficult to maintain that life is common in the galaxy. In the past the only evidence for life has been on earth. The spacecraft investigations of our solar system, especially of Mars, revealed no other life.

Meteorites have been dated at about 4 billion years, which shows they originated in the early days of the solar system. Most meteorites are of either nickel-iron or stone, and there is no indication that they are other than random pieces of matter left over from the days of formation. There is, however, a very rare type of meteorite called a carbonaceous chondrite which is black and quite soft (in the sense that it can be easily crumbled). There is a tiny percentage of water in it, and usually there are some carbon compounds. Very few such meteorites survive the fall to earth, but some do. About 25 have been found. Studies of these have been made, and some surprising results have come to light. Several meteorites contained elements of what we consider to be the building blocks of life: fatty acids, amino acids, and longer chains of carbon and hydrogen atoms. Additional studies of meteorites in Antarctica where they strike earth and lie frozen in an almost pristine state have also shown evidence of compounds which have formed in the direction of life.

The data accumulated on earth point to the origin of life only half a billion years after the oceans were formed. The first examples were the simplest chains of molecules, followed by simple compounds, followed by cells. The dominant forms of life for 2 billion years were probably bacteria cells and blue-green algae (cyanobacteria). Earliest life probably contained molecules of either DNA or a closely related and equivalent substance. The DNA most likely appeared in the form of loose plasmids, rather than organized chromosomes. Early life did not contain the nucleus found in modern cells.

The studies of meteorites can be interpreted two ways: the chains of molecules leading to life might have arrived here by meteorite or from comets, as has also been suggested, and the rest of the process occurred naturally in the primordial earth environment; or the evidence might simply show that atoms tend to come together naturally to form compounds in the direction of life whenever they are given a chance to do so. Whether that chance comes on a planet or out in cold space aboard a meteorite is not necessarily important. Life could be common in the universe wherever the right conditions existed on a planet, and those who hold to the anthropic principle that we are alone would be wrong.

The right conditions existed on earth because along the way the earth survived a major planetary crisis. Sunlight trapped in the earth's dense early atmosphere could not escape, and there was a severe greenhouse effect. The introduction of oxygen, first by photolysis of water vapor and later by photosynthesis, lowered the temperature, in time, preventing what might be described as a "runaway greenhouse effect," which is self-stimulating and self-amplifying. Had a runaway effect happened, a dense, thick atmosphere and a high-temperature surface might have developed on earth, as it did on Venus.

A second major crisis occurred about 1.8 billion years ago when massive glaciation spread down from the pole to cover 10 percent of the surface of the earth. Some estimates indicate that the earth was within one tenth of a degree F in average temperature from becoming a planet too glaciated for life to continue.

While modern theories suggest that life may be common, they also show that it may be rare and that, therefore, a habitable planet may be rare. Had the earth been only a tiny distance farther from the sun, runaway glaciation would have occurred; had it been only marginally closer to the sun, a runaway greenhouse effect would have been likely. The conclusion is that the habitable zone around a star like the sun is extremely narrow—0.95 to 1.01 au. Thus, a variation of 5 percent less or 1 percent more, and the earth would not be as we know it.

All of the statistical analysis about possible habitable planets does not tell us what we would really like to know. How many are nearby? Are these basically "earth-like"? If there are 880,000 habitable planets in the Milky Way, it does not necessarily follow that any of them are close to us. Statistics can be misleading. Which of the 200 stars close by has a habitable planet—one very close, one in the middle distance, or one so far that we could not possibly travel there with any of the suggested propulsion systems?

Another problem arises in applying statistics. There are far more stars near the center of the galaxy than there are out where the sun lies, two thirds of the way to the edge of the galactic spiral. Habitable planets would presumably be closer together there, on the average. What we need to know is, how "average" is our section of the galaxy?

Within about 20 light years of the earth are 111 stars. The majority of them are normal stars. A hundred are stars we can see, and 11 more have been detected by other means. Most of the 11 are small and circle larger stars nearby; they have been found by observation of gravitational effects of one body on the other. For various reasons, 68 of the stars are not suitable: 3 are too massive and would have short lives (Sirius, which appears as the brightest star in the sky, is one of these); 7 are stars that have collapsed into white dwarfs and would have destroyed any planets with their death throes; 57 are simply too small for a suitable planet to have existed long enough to become habitable. One star, otherwise suitable, has a white-dwarf companion whose gravitational influence would be disturbing to planetary formation.

This leaves 43 stars as candidates for our starship trips, which would seem to be a very substantial group. Unfortunately, too many of these are strictly borderline prospects. Some are in multiple systems that would require a very complex orbit for a habitable planet. Twenty-eight are at the lower limit of size, so that a habitable planet around one of them would be unlikely. Planets near small stars are stopped from rotating quickly by tidal effects. A large moon would mitigate that problem, but the statistical probability is slight that a given small star would have a planet at the right distance for heat as well as a moon just the right size to counter rotational problems. Slow-rotating planets are not conducive to life.* In our own solar system, only one planet has a single moon, with the exception of Pluto which is a special case.

The term habitable planet does not necessarily mean the planet must be earthlike; it means it must be one on which earth-type life could with some luck exist and flourish. Many of them might be very earthlike. They would be about the same mass as earth, on the average a little smaller, and there would be a familiar atmosphere of nitrogen and oxygen with other gases thrown in. The day-night cycle would be reasonably near 24 hours.

The seasons would be more or less like *ours* depending on the exact nature of the orbit around the star and the inclination of the planet on its axis. There would be oceans and rivers, deserts and mountains, and polar ice caps. Most of what we experience as weather would be duplicated on it.

But there are other types of planets, not as much like ours, that might be habitable. One could be a satellite of a gas giant like Jupiter, both of them revolving around their sun. A planet such as that would have odd cycles of day and night, but that would not necessarily preclude habitation. It would have a total eclipse every day, with

*Some authorities disagree that slow-rotating or tidally locked planets would be poor prospects for life. A planet could avoid freezing its atmosphere or oceans on the dark side by means of heat transport via atmosphere circulation; this process occurs on Venus, where the day-side and night-side temperatures are the same.

the shadow of the gas giant falling entirely over the smaller planet. The nights would be dominated by the great "moon" in the sky.

Planets that are binary—twin planets—are potential homes for mankind. The two would revolve around a common center of mass. Habitable planets can exist in some doublestar systems or even multiple systems. The orbits must be stable, and the stars must have a combined radiation which would allow reasonable temperatures on the planet's surface. This could include two-star systems in which the stars were fairly distantly separated. Either or both of the stars could have a habitable planet. If the stars were extremely close together, a planet could orbit them both, but at a considerable distance.*

There could be planets in which only a narrow belt would be habitable. A world with a high equatorial inclination might have a belt between latitudes 14°N and 14°S where the conditions are good for mankind. Outside the 2000-mile-wide zone, it would be too cold in winter and too hot in summer. Another type might have mild conditions only in regions around the poles. A world could have a ring, as Saturn does. Rings around planets may be much more common than has been supposed for centuries. The discovery of a set of rings around Uranus in 1978 and 1979—9 in all—bears this out, as well as the discovery of a ring around Jupiter. Rings around earth-size planets, however, might be thinner than those around large planets.†

There could be planets with much more ocean than earth has and planets with much less. Mars has the same amount of land mass as earth, though it is smaller, because none of it is covered with ocean. It is possible that the larger planets within the range of habitability would

*There is some evidence that except in very close binaries and very widely separated binaries, planets are unable to form at all.

†It has been theorized recently that the earth had a small ring at one time.

be more oceanic and the smaller planets would be mostly dry land.

There is one further, rather interesting chance. In 1962, Harlow Shapley published a paper in which he suggested that life might exist without the benefit of a sun on bodies intermediate between stars and planets. There might be, Shapley thought, "myriads of these bodies which are not orbitally obedient to any star—these in addition to the planets of all sizes that are immediately subservient to stars."

He pointed out that a sufficiently large body would generate its own internal heat, but the crust would be solid, and water could exist in liquid form. On earth, the same kind of internal heat exists, emanating from volcanoes and hot springs. On the surface of one of the dark planets the landscape would glow in the deep infrared. The glow could be utilized by organisms for photosynthesis under certain conditions, and life of some form could evolve. But a dark planet would have a very different biology, and the gravity likely from such a large planetary mass would probably preclude human habitation.

The dark planets would be found in the regions between the stars; there might be one close to earth. Pluto is too small to be responsible for the gravitational disturbances that have been observed in the orbit of Neptune. Perhaps a large, dark planet exists somewhere on the edge of the solar system, out beyond Pluto's orbit. There might be several between the sun and the nearby stars. At present there is no way of detecting one.

If the borderline cases and those statistically improbable are eliminated from the closest 111 stars, we are left with 14 stars that might have a chance better than 1 in 100 of having a planet where we could survive. These 14 are: Alpha Centauri's system of 3, Epsilon Eridani, Tau Ceti, 70 Ophiuchi A, Eta Cassiopeiae A, Sigma Draconis, the 2 stars of 36 Ophiuchi, a star called HR 7703 A, Delta

Pavonia, 82 Eridani, Beta Hydri, and HR 8832. In the total list, there is about a 3-to-2 probability of finding a single habitable planet—3 to 2 against.

So far, the reasoning about destination has followed along the lines suggested by William Whewell so long ago: find the stars that are most like the sun and you will find the planets most like the earth. But he also speculated that chances would be improved by finding even one planet going around another star—not a statistical planet but one observed and recorded by telescope, one about which something concrete and physical could be determined. As he pointed out, if there is one real planet going around one real star like the sun, then the supposition that planetary systems are common in the galaxy is not just an extrapolation of the solar system we live in. Such a discovery would have profound significance and would tend to confirm the statistical projections all the way from nearby space to the average of potentially habitable planets in the galaxy.

The most difficult of the problems facing astronomers in trying to detect a planet circling another star is the difference in brightness between the star and the planet. If a planet the size of Jupiter orbited a star like the sun at the same distance it does in our solar system, it could be detected at 10 light years, provided the separation was wide enough and the telescope was powerful enough to detect that separation. But a planet the size of Jupiter would be so faint, shining solely from reflected sunlight, that it would not be visible unless it were only *half* a light year away. Finding a planet as small as the earth would be impossible; it could not be detected unless it were closer than a trillion miles. In either case, a large or small planet would show up only as a tiny speck of light. Nothing could be discovered about the planet, for the same reason that almost every bit of data about Pluto is still followed by a question mark in most textbooks.

A large telescope on the far side of the moon might

detect a planet going around a very close star by direct means, but no such telescope exists. Fortunately, a different method has been found. In 1844, Bessel detected the companion star to Sirius by nonvisual means, through calculations based on the effect its gravity had on a larger star. This displacement wobble in the sky was an indicator that some object, even though it could not be seen in a telescope, was in orbit around the visible object.

Bessel's method would also work for a planet going around another star, but the wobble would be extremely tiny. The question, of course, is whether the wobble would be large enough to be detected in even the closest star. Viewed from a distance, the sun would exhibit a small displacement caused by Jupiter, and the displacement would reflect a 12-year cycle, the time it takes Jupiter to revolve once around the sun.

Such an assessment of Alpha Centauri was tried, but there were reasons why the method did not produce the desired results. It is difficult to detect a wobble in the motion of one member of a binary-star pair because the motion of such a binary is not known as precisely as the straight-line motion of a single star across the sky. Alpha Centauri was not only a binary system; it was a multiple system, although the third star, Proxima, was too far from the main pair to have much gravitational effect.

The next closest star was Barnard's Star, at 5.9 light years. This small, red-dwarf star is only 15 percent as massive as the sun. Red dwarfs have a low surface temperature, and they are among the longest-lived stars in the galaxy. They are also faint because of low energy output. It is the extremely low energy output which gives them their longevity. About 80 percent of the stars near the sun are red dwarfs.

If a planet were detected orbiting Barnard's Star, it would be particularly important since Barnard's as a red dwarf, is among the most numerous types in the galaxy. The discovery would indicate that most main-sequence stars have planets and therefore the statistical analysis for habitable planets was valid.

Observations were begun in 1937 by Peter Van De Kamp at the Sproul Observatory in Pennsylvania. By painstakingly taking a massive number of photographs of Barnard's Star every six months (2413 plates taken on 609 nights) from opposite sides of the earth's orbit, he was able to detect what he thought to be a wobble in the star's position, which indicated the presence of a massive companion object. By 1956, he was reasonably sure that the companion object was a planet about the size of Jupiter, but only 2.71 au from the star. (Jupiter is about 5 au from the sun.) The planet had an apparent orbit of about 11.5 years, nearly the same as that of Jupiter. On April 8, 1963, Van De Kamp announced his results. In 1969, further refinements of his data indicated there was another planet at about 4.17 au with a 22-year orbit. Saturn goes around the sun in 29.5 years at a distance of 9 to 10 au.

Some astronomers felt that Van De Kamps's observations were simply too close to the limits of his instrument and too subject to error, but the results have been tentatively accepted.* In 1973, others suggested that Van De Kamp's data showed there were three planets, the largest one bigger than Jupiter and orbiting at the average distance of Mars from the sun. The other two were somewhat similar to Saturn and Uranus, with the most distant 4.5 au from Barnard's Star.

In 1974, Van De Kamp thought he had detected a very large planet—6 times the mass of Jupiter—orbiting Epsilon Eridani. For 61 Cygni, which is 11.2 light years away, a companion has also been theorized on the basis of measurements of slight gravitational wobble. In this case the companion has been estimated at 8 times the mass of

*Eichorn in the mid-1970s failed to find clear evidence for any companions. Many others have emphasized that the data are inadequate for a definite conclusion.

Jupiter. Ross 614, a small red-dwarf star like Barnard's at 13.1 light years, may have a companion that is 8 percent of the mass of the sun. The companion would be another "star," since it would be massive enough for the fusion process, though it would shine very dimly. Another large companion about 7 times the mass of Jupiter may circle Lalande 21185 at 8.1 light years.*

Using photography to find planets is a laborious process requiring years to accumulate a sufficient number of observations to support a conclusion. As many years are needed for confirmation as the time taken for a massive object to revolve around the star. The use of earth-based telescopes hampers observation because the boiling atmosphere makes the images on the photographs blurred and hard to measure. When the measurement for 61 Cygni was made, the blurred and highly enlarged photographic image was a sprawling amoeba covering a space many times the estimated ability of the telescope to separate faint objects. Only by using sets of extremely fine cross hairs and measuring centers of the images by averaging was the wobble motion detected at all.

The telescope used by Van De Kamp was a nineteenth-century instrument of exquisite manufacture but not necessarily the best one for the purpose. In 1964, a special "astrometric" telescope went into operation at Flagstaff, Arizona, at the observatory where Pluto was discovered and where the famous "canals" of Mars were so often studied and mistakenly assumed to be real. By 1984, it could confirm the existence of Saturn-sized planets around nearby stars.

It should be possible to use the transit of a companion body to confirm its existence. A dark companion crossing in front of a star as viewed from earth would show a slight change in the star's spectrum. There would be a

*All of the other reported planetary companions are highly suspect because of the minute measurements and subjective interpretation of the data.

shift toward the blue as the planet crossed the limb of the star, followed by an abnormal reddening at the center of the transit. As the planet approached the other limb, there would be another shift to the blue. The late Frank Rosenblatt of Cornell University proposed that this transit "signature" could be observed by a system of three wide-field telescopes at widely separated sites connected to a single computer. He suggested that one planet per year could be discovered by this means.

If a telescope were in space beyond the atmosphere, it could resolve to the theoretical limits determined by mirror size. When Space Telescope becomes operational in the late 1980s, one of its first tasks should be to look at the nearby stars for confirmation of planetary bodies and for new possibilities. There are also new techniques such as speckle interferometry, which, if used in space, would also help detect planets around other stars.

Estimates of the number of habitable planets in the Milky Way are what kept so many people believing in UFOs, engaging in the search for extraterrestrial intelligence (SETI), trying to establish active communication with alien intelligences, and other seemingly thin dreams. The reasoning, of course, is obvious. If there are 880,000 planets in the galaxy which could be described as "earthlike," then there are a predictable number on which intelligent life has evolved with civilizations of high technology. Making some tenuous assumptions, we can estimate the number of planets on which some sort of animal life has evolved and the number of these on which life has risen in intelligence to the point where there is a civilization.

A mathematical excursion into attempting to identify the number of civilizations in the galaxy is the famous $N=R^*f_p n_e f_l f_i f_c L$. This relatively complex-looking but harmless equation can show how many civilizations might exist in our galaxy right now and how long their mean (mathematical) lifetime has been. It has become a symbol to those who believe in interstellar communication.

The original formula was developed by Frank Drake of the Cornell Center for Radiophysics and Space Research. N stands for the number of currently existing civilizations in the galaxy; R^*, the rate of star formation averaged over the existence of the galaxy in units of the number of stars per year; f_p, the percentage of stars with planetary systems; n_e, the mean number of habitable planets; f_l, the percentage of those planets where life has already started; f_i, the planets where the life has become intelligent; f_c, the percentage of planets where life has reached a technological stage; L, the mean lifetime of those technological civilizations.

Statistically, at least, there are possibly other civilizations of a high technological order somewhere in the Milky Way Galaxy. But civilizations rise and fall; some destroy each other; some will, no doubt, have destroyed themselves. The earth has been in existence almost half the time it can support life, and it has developed only one civilization—accounts of Atlantis and the antediluvian world aside.

Speculations about other civilizations led to the romantic, if not impractical, Project Ozma (after the princess in Oz), the first attempt to listen in for evidence of another civilization among the stars. The listening began at 4 A.M. on April 8, 1960. It stopped, after a total of 150 hours of listening, that July. There are basically two types of extraterrestrial signals which might be detected from earth accidentally: those which leak from various planet-based electronic systems (the earth has been radiating broadcast electronics into space for most of this century) and those which are actual interceptions of an existing interstellar communications link. Ozma studied 2 nearby stars: Epsilon Eridani and Tau Ceti, both about 12 light years away.

To find communication and broadcast leakage anywhere within 100 light years of earth would require an antenna much larger than anything available now—a few square miles of antenna-effective area. The chances of intercepting a random interstellar communications system are extremely small for a variety of reasons, not the least of which is whether we would recognize one if we found it.

Despite the potential difficulties, real and theoretical, Project Ozma was followed by other attempts to receive and initiate interstellar communication. In 1968, the Radio Astronomy Station of the Research Institute of Radiophysics in the USSR was used to study 12 stars. Additional experiments using the same station, and others, in the Soviet Union were conducted in 1970. Between 1970 and 1976 seven other studies were carried out to detect extraterrestrial communications, including one in 1975 by F. D. Drake and Carl Sagan using the 1000-foot radio dish at Arecibo, Puerto Rico.

Although it was never built, Project Cyclops, designed in 1971, could have detected leakage signals similar to those emitted from earth at a distance of 100 light years, and it could have detected an omnidirectional radio beacon at a distance of 1000 light years. In 1978, the Arecibo dish was again used, this time focusing on 200 nearby stars similar to the sun. Despite all of this effort, no artificial signals have ever been detected. Efforts are still continuing, however, at several locations around the world, and it has been proposed that a very large antenna be sent into space aboard the space shuttle to be used, part of the time, in seeking extraterrestrial intelligence.

The reason no response is coming from the possible civilizations out there, and no evidence has been received that they exist, may be the distance. If the probability figures for civilizations are anywhere near accurate, on the average there might be 1 civilization within a distance of 500 to 600 light years. The correct figure might be 1 in 1000 light years, or 1 in 10,000. It might be zero.

STARPROBE

"Oh ye heavens, of your boundless nights, Rush of Suns and roll of systems . . ."

TENNYSON

With the odds a relatively poor 3 to 2 *against* finding a habitable planet among the nearby 14 candidate stars, it is important to find out if those odds can be improved. We have come close to exhausting mathematical models; what we need is more real data. There are three ways we can acquire additional facts for mankind's first starflight: by refining the information about planetary formation and evolution in our own solar system; by beginning a concentrated search for other planetary systems using both earth-based and space-based telescopes; and by sending unmanned spacecraft—starprobes—into interstellar space and to other stellar systems.

Although the mathematical models may be inaccurate because of our lack of information, they can be used as a starting place for new programs of study. How accurate, for example, are our currently held theories of planetary formation and evolution which were used as a basis of predictions for planets around other stars?

Mercury has a magnetic field. Venus, our sister planet in size and density, does not. Why should a tiny planet, the smallest in the solar system, have a magnetic field and Venus none? We would like to know why Mercury formed with a heavy iron core when our prespacecraft theories would not have predicted one. Why was Mercury much "hotter" in our theories than we discovered it to be in fact? Mercury is confusing: it is like the earth on the inside—the core, the magnetic field—and it is like the moon on the outside. What caused the lobate sinuous scarps several hundred miles long and 800 feet high in the Mercurian highlands?

Are our suppositions about the evolutions of Venus, earth, and Mars correct? Is what we see now from our spacecraft and landers only a result of one planet's being too close to the sun, one too far away, and one just right? Are the conditions so pleasant on earth only because our planet represents a cosmic "Baby Bear's porridge"? Why does Mars have what seem to be 50,000-year cycles in which the conditions are altered substantially? Mars had running water once. Why not now? Will it in the future? Would we be inclined to call Mars a "habitable planet" by our current definitions when the 50,000-year cycle it's in now ceases and the conditions are more favorable to life?

The list is endless. Galileo's moons of Jupiter were a series of surprises. Titan, Saturn's big moon, seems to have been formed in much the same way as other large moons in the solar system, but it has an atmosphere of very respectable proportions—denser than Mars, a surface pressure greater than that of earth. And the conditions on the big moon seem to resemble, more than anything else, a primordial earth before our early atmosphere began reducing to the one we have now. What is Uranus like? Neptune? Is Triton, the big moon of Neptune, like Titan, or is it a dead, frozen world which has remained the same since our solar system was formed? We think we have discovered a "moon" circling an asteroid. Is that common out in the asteroid belt, and if so, how do we account for such a situation in our theories of how the solar system was formed?

The solar system we have seen from our unmanned spacecraft is not the one we always thought it was. We must postulate a new theory which takes into account all of the variations in planets and moons that we have found. It was so much easier fifty years ago when dead planets remained dead, sterile moons remained sterile, and nature didn't come forth to confound us with Her diversity. "God does not play dice with the Universe," Einstein said. From current evidence within the solar system, He must have played Pickup Stix.

Since we still believe that the best place to find a habitable planet is around a sunlike star, we can begin the long road to starflight by finding out as much as possible about our own solar system. That means we must schedule additional fly-by missions to Mercury, perhaps even a small lander like the old *Surveyor* which landed on the moon in the 1960s or a Viking lander craft similar to that still on Mars. We must perfect synthetic-aperture radar-equipped orbiters to pierce the clouds and secrets of Venus and construct armored landers to determine the surface conditions. Above all we must again land on Mars with vehicles and instruments to investigate and puzzle over the Red Planet and see how it fits into the picture of the solar system we have sketched in theory.

Beyond Mars, in the outer solar system, we will examine asteroids, comets and meteors, and interplanetary dust. We will land on the Galilean planets and find out why one moon of Jupiter is a frozen cameo of the early days of the solar system and another, not a great distance away, is a violently active, volcanic world as unlike a "moon" as anything we could have imagined. We will see the green hues of Uranus, the blues of Neptune, and the surface of Triton. We may be able to find out why Pluto is unlike all the other planets in the solar system.

When all of this planetary research is digested, we may discover that our theories of evolution in the sun's family of planets were substantially incorrect. Our mathematical models of what might happen in another solar system might also prove to be incorrect. The odds for finding habitable planets may be much improved. Fortunately for our hopes for starflight, the next twenty years will bring in a large amount of new information. A mission has been planned by the European Space Agency to intercept Halley's Comet on its return in 1986. VOIR, an acronym for Venus Orbiter Imaging Radar, has been funded by NASA to investigate the secrets of the second planet from the sun. *Galileo,* a spacecraft designed to orbit Jupiter, will go toward the giant planet in the 1980s. The *Voyager* spacecraft which provided so much information from Jupiter and Saturn encounters will continue onward into the depths of the solar system to Uranus in 1986 and Neptune in 1989. The ending years of this century may see new explorations of Mars and Mercury and a major landing on Venus.

While all of that is being done, we will be applying our technologies to solving the question of whether there are planets circling the nearby stars and we will develop new techniques which can determine something about the planets. The currently used methods are hopelessly inadequate for confirming the existence of planets around other stars, since none of the present instruments was originally designed to meet the requirements of ultra-high precision and long-term stability necessary. We have spent over 40 years compiling planetary searches on a total of 14 stars, and the most generous conclusion that can be drawn from the data is that 4 of the stars studied *may* have planets. The least generous interpretation is that the data are entirely inconclusive: we don't know if there are *any* planets around the nearby stars, let alone habitable worlds.

In the spring of 1976, several workshops which dealt with the problem of detecting extrasolar planets were held under the auspices of NASA. One of these was Project Orion, quite different from the design study of nuclear-pulse engines with the same name. A principal objective of the study was the design of programs to improve ground-based astrometric telescopes. One of the suggestions made by the study was the building of an instrument called the "imaging stellar interferometer," the ISI.

Among the potential sites for the instrument were Black Mesa, Arizona; Sierra de los Filabres, Spain; Mauna Kea, Hawaii; and the Guadaloupe Islands. The ISI could detect planets orbiting stars at distances of 10–40 parsecs. Studying stars at that distance, it could search a sufficiently large and homogeneous sample to provide reliable statistics on the nature of planetary systems. Whether the ISI is ever built is a matter of funding and further design study. The conclusion of the Orion Project scientists was that current methods of extrasolar planetary detection could be improved by more than an order of magnitude. The cost, it was estimated, might be about the same as a low-budget planetary mission in the solar system (perhaps $100 million), even if the effort included placing telescopes in earth orbit.

Other studies have proposed that special occulting

disks placed over conventional telescopes would be a vast improvement in extrasolar planetary work.* Finding an earth-size planet by using an occulting disk would be difficult, however. Even with the device attached to a telescope, it would take one with a diameter of 1200 inches to detect an earth-size planet at 15 light years. Unfortunately, a 1200-inch telescope cannot be built on earth; 300 inches is probably the size limitation for an earth-based single-mirror telescope. If it were larger than that, gravity would deflect the mirror surface and the instrument would be optically useless. Multiple-mirror systems that employ smaller individual mirrors acting in concert are practical, and one has been put into operation at Kitt Peak National Observatory in Arizona. With some of the latest multimirror systems it might be possible to build a very large earth-based telescope that could directly detect planets circling other stars.

By far the most advanced studies aimed at finding extrasolar planets have projected the placement of a large telescope in orbit around the earth. Positioned high above the boiling atmosphere which hampers so much earth-based observation, it would use a variety of methods to look for planets in a candidate list of more than 100 stars. In space, mirrors are not troubled by gravity and do not deform, which means that telescopes with single mirrors could be much larger than those on earth today. Multimirror telescopes in space could be truly gigantic. Proposals have been advanced for space telescopes as large as 5000 inches in effective diameter. Gerard O'Neill suggested a telescope with 200 glass seg-

*If a star at 5 parsecs, 16.5 light years, has a Jupiter-like planet orbiting at 5 au, the planet could easily be detected with modern telescopes if the problem were one of separation from the star alone. Unfortunately a planet is much fainter than a star. Jupiter is 23 magnitudes fainter than the sun. The problem thus becomes one of detecting two objects of vastly different brightness. For direct visual detection in the above case, a 1200-inch telescope would be needed; using an occulting disk to mask the brighter star in the telescope would improve detection. A 200-inch telescope should theoretically be able to detect the Jupiter-like planet.

ments, each about 3 feet in diameter, in a multimirror format which would be carried to earth orbit in several trips of the space shuttle.

Electronic systems have been used to enhance the performance of telescopes on earth during the last thirty or forty years. In space, with developments which will come from the technology around the space shuttle and sophisticated satellites, it will be possible to equip a telescope with instruments far beyond the capabilities of those in use today. One of the problems with the data on other solar systems now is that they come from instruments of doubtful stability and accuracy. Any study of extrasolar planets with telescopes requires tremendous accuracy and extremely tiny measurements. Electronic systems have been developed with military applications partly in mind which can achieve accuracy equivalent to making a hole-in-one on a golf course in space from the distance of the earth.

Aboard a space-based observatory it will be possible to use instruments which cannot be used on earth. Infrared observations from the earth's surface are almost impossible because the water vapor and carbon dioxide in our atmosphere are very efficient absorbers of radiation over many regions of the infrared. A telescope in space which could use infrared techniques for planetary search has been studied in the last few years by Stanford University, Hewlett-Packard, and Lockheed. Infrared detectors have been tested in the Kuiper Airborne Observatory—the same observatory from which the rings around Uranus, so difficult to see from earth-based instruments, were discovered.

Space-located multiple mirror instruments could detect an earth-size planet orbiting Alpha Centauri, and special techniques using different wavelengths of light could find earth-size planets as distant as Tau Ceti. Special eclipse-detection methods have been proposed that might locate one planet of the Jupiter size per year as a computer-controlled series of three telescopes checked 9000 stars each night.

Whether any of the proposals for giant space telescopes will be accepted for future implementation is questionable. Eventually there will be a highly developed astronomical center in space, but it is years away from reality. Meanwhile, there is funding for a space telescope to be flown aboard a space shuttle in the 1980s. With the currently accepted design of the space telescope (called the LST, for large space telescope), planets of the Jupiter size could be detected at a distance of 10 parsecs in less than 20 minutes using an area photometer. The space telescope may provide the first hard information about planets in other solar systems. By the end of this century, the data on potential target stars should be vastly improved.

About forty years ago Einstein published a small note in the magazine *Science* on focusing starlight by the gravitational field of another star. Current technology and technological trends suggest that it may be possible someday to use gravitational focusing of light and other radiation in the electromagnetic spectrum for directional observations. The idea also has applications for communications over interstellar distances. The prospect involves highly sophisticated spacecraft, at 2200 au from the sun, that would use the gravitational field of the sun as a spherical lens to magnify the intensity of light from a distant source. Von R. Eshleman of the Center for Radar Astronomy at Stanford discussed the idea briefly in a letter to *Science* in 1979.

There have been suggestions that such a gravitational lens system be used to investigate the spectrum of planets around nearby stars. If this can be put into operation sometime in the next century, we may know a great deal about the planets going around the nearby stars without ever visiting them. It should at least be possible to identify the best target stars for a space probe. As an unmanned vehicle, it could look at and evaluate star systems with improved versions of the instruments used by the successful Mariner, Viking, and Voyager spacecraft in our own solar system. Thus starprobes can be used to survey the most likely systems for an earth-like planet long before a manned ship ever leaves the solar system.

How soon a starprobe might be built is an open question. It does not appear that the gas-core fission engine will ever be developed. This leaves fusion as the best choice, but it is not now available except in the nuclear-pulse design. The latter, of course, is not under development for a variety of reasons, a major one of which is that it is illegal under the nuclear-limitation treaties.

Any starprobe worthy of the name is utterly dependent on the development of propulsion systems—fusion being the one most favored now. But since no one has produced *net* energy from *any* kind of fusion, how long will it be before there is a working fusion engine for a starprobe? In 1972 the Rocketdyne proposal for the space-shuttle main engine was a stack of documents four feet high. The proposal was complete with a design for a prototype full-scale thrust chamber that had been fired successfully on the test stand. Yet the space shuttle still had not flown by 1980, and there were still problems with the main engines, though Rocketdyne had benefited from 30 years of experimentation with liquid-fuel rocket engines of basically similar design and a spaceship powered by them had already reached the moon. Only truly astonishing advances in the state of propulsion art can keep an effective circle of limitation from occurring—a circle limited by propulsion, fuel loads, and engine design.

Ideally, the information from a starprobe should be delivered to earth within the lifetime of the scientists who designed the mission. They would be the best equipped to evaluate the findings. Even if the fusion engines we hope for are developed, information returns of 50 years or so will limit the circle of exploration to the closest stars. Still, within that circle drawn by our hoped-for technologies are the two Alpha Centauri stars, Barnard's star, Epsilon Eridani, and Tau Ceti. Barnard's star is particularly interesting because it represents a class of stars of which there are 9 others nearby. If a starprobe were sent there and found a planet even marginally suitable for mankind, the other 9 would be considered similar and would be expected to have similar planets.

In addition to the candidates above, there are other systems of great scientific interest but little chance of having a planet suitable for mankind. These are Wolf 359, Lalande 21185, Sirius, Luyten 726-8, Ross 154, Ross 248, Luyten 789-6, Ross 128, 61 Cygni, Epsilon Indi, Procyon, E2398, Groombridge 34, and Lacaille 9352.

Wolf 359 is a small red-dwarf star. If, as our tentative data indicates, there are one or more planets circling Barnard's star, Wolf 359, which is similar, may have one or more also. Resembling these two are Lalande 21185, all three Ross stars, both Luytens, Lacaille 9352, the Groombridge star, and E2398. Epsilon Indi and 61 Cygni are both small stars that are unlikely to have planets. The others are too hot to have the same kind of worlds our solar system has, if our understanding of stellar and planetary evolutions is correct.

Could a starprobe discover the existence of a civilization on a planet circling another star? Certainly, but the probe itself could not tell us much unless its mission were lengthened. It would be launched from our solar system to reach as high a velocity as possible to cut down the information-return time. For fusion-powered probes this time leaves only a one-way, full-speed run. It would enter another stellar system at very high velocity (a 0.14*c* starprobe to Barnard's Star would cross the entire system in about 10 hours), making detailed observations difficult. Decelerating the probe to enter at slower speeds would considerably lengthen trip times, making them roughly twice as long.

The solution is to design a starprobe large enough to hold several small planetary probes of an advanced Voyager type. These would investigate the star system, and some could be designed to take up orbit around the most interesting planets. A highly sophisticated probe around an earthlike planet could detect the remains of a civilization much in the same way NASA used radar to "discover" long-unseen Mayan canals.

NASA developed a radar system for the military in which the resolution for distinguished objects and patterns was as small as 45 feet. The ruins of Mayan irrigation canals were discovered on images gathered by the new radar during an early test from aircraft in 1977–78. The device, called "synthetic-aperture radar," was later used in a 1978 oceanographic-research satellite. Though the buried Mayan canals were only a few feet deep and about 10 feet wide, they were mapped during a survey of 50,000 square miles of Guatemalan and Belizean tropical jungle.

A synthetic-aperture radar aboard a starprobe planetary-exploration orbiter could detect an existing civilization or the ruins of an old one on any planet among the nearby stellar systems.* Such a discovery would unquestionably generate immense interest in a manned starship, and it is partly for this reason that a starprobe is necessary. Any major discovery of habitable planets or life of any sort would result in a tremendous push to build a starship.

Around 1970, Krafft Ehnicke, who was then the chief scientific adviser for the Advanced Program Division of North American Rockwell's Space Division in California, coined names for the various zones of space around the

*Synthetic-aperture radar would be chosen only for a planet covered with thick clouds or jungles. A relatively open planet could be better investigated by conventional optical means from orbit.

sun in order of their increasing distance. The first was the stellar magnetosphere, followed by the circumstellar zone, stellar gravisphere, suboptical zone (SOZ), quasioptical zone (QOZ), and the zone of isolation (ZI). The first two he lumped together with the term "ultraplanetary space" and proposed that probes be sent there to determine the conditions. He defined ultraplanetary space as that area out to a distance of 0.1 light years.

The exploration of the ultraplanetary space around the sun must proceed before a starprobe can be sent out, principally to gain knowledge so that proper protection can be designed for the starprobe. An ultraplanetary vehicle would serve much the same purpose as the early missions to Jupiter and Saturn, which showed that the asteroid belt could easily be navigated and that the rings of Saturn could be approached without damage to future spacecraft.

To a degree, this sort of exploration has already begun. *Pioneer 10* and *11* and *Voyager 1* and *2* will explore beyond what we usually refer to as the solar system before their distance becomes so great that the communication is terminated or the spacecraft run out of power.* But it would be useful to send at least one ultraplanetary probe out to 0.1 to 0.5 light years. It can be assumed that the nature of the space around the sun is much the same for all stars and indicative of what would be found around another system. True interstellar space is basically empty with a few molecules of hydrogen and other assorted elements every few cubic inches and relatively nondangerous to penetrating spacecraft.

An ultraplanetary probe could be launched from near earth orbit after a delivery by a future space-shuttle-type

*Pioneer 10, now 2.3 billion miles out, has shown that the sun's atmosphere and magnetic envelope extends out an enormous distance. The implication is that if you were living on any other planet, you would find a solar environment around it like that surrounding earth. All sun-like stars would be basically the same and so, then, might their planets.

vehicle. Using "conventional" propulsion and a complex gravity-assist technique, the ultraplanetary probe would explore the regions outward from the sun over 20 years.

The gravity-assist method of adding velocity to a spacecraft was first tried successfully with *Mariner 10* to Mercury in 1973. It was also used by *Pioneer 11* to go to Saturn and by *Voyager 1* and *2.* The spacecraft uses the gravity of one planet to assist in building up the required velocity to visit the next. If more than one assist is used, the velocity increases each time, and the larger the planet, the greater the increase. It is for this reason that Jupiter is the preferred planet.

The sun can also supply gravity-assist, being the most massive object around. The procedure is unnecessary for planetary flights within the solar system, but an ultraplanetary probe would gain a healthy boost.

Hermann Oberth, a German spaceflight pioneer, showed as early as 1929 that the highest velocity for escape from the solar system is obtained from a steep descent orbit toward the sun and a high-propulsion impulse during the perihelion maneuver—a prospect which has appeared in everything from *Amazing Stories* to an episode of *Star Trek.*

The most complex of the solar-system maneuvers using the sun's gravity is a launch toward Saturn, a rebound toward Jupiter, a powered approach very close to Jupiter for an additional boost (known as a "peri-Jove maneuver"), and a steep descent orbit toward the sun. An ultraplanetary probe able to accelerate at only 0.1 g for three days toward the sun during its perihelion maneuver would reach a speed of 372 miles per second, or 1,339,200 mph.

A hypothetical ultraplanetary probe, *Prometheus,* could cover 100 au in only a year. In 10 years, it would be out 1000 au, 3 times farther than *Voyager 2* will be in 100 years. Within the 50-year limit of information return, the probe would have covered the space around the region of the sun out to 0.1 light years. It would not be a real starprobe, however. It would take until A.D. 4140 to reach Alpha Centauri.

Among the bits of information brought to earth would be the percentage of molecular, atomic, and ionized hydrogen, helium, and oxygen in the interstellar medium. *Prometheus* would gather data on how the medium varies with distance from the sun. This information would be important in designing a true starprobe and especially important in deciding how much shielding to specify for a manned starship.

While ultraplanetary probes were reporting their findings, the construction of the first true starprobe could begin in the orbits high above the earth. One design has already been completed by the British Interplanetary Society (BIS). While parts of their Project Daedalus are not practical in terms of what is called current technology (propulsion, for example), most of their concept is a reasonable extrapolation of present technological trends. A study group for the starprobe was established on January 10, 1973, and BIS released its final report on the design in 1978.

All of the Project Daedalus personnel were professional engineers and scientists. They included members of the British Aircraft Corporation, Atomic Energy Authority, European Space Agency (ESA), Royal Air Force, Rocket Propulsion Establishment, and other scientific facilities in Europe. Although they were not acting in their official capacities while studying starflight, their results were the product of a professional and concentrated approach. "Final Report of Project Daedalus" involved more than a dozen people working for 5 years—10,000 man hours.

The *Daedalus* starprobe would require 20 years of design, manufacture, and vehicle checkout. That estimate is probably a bare minimum. The mission proposed by the British Interplanetary Society would use nuclear-pulse engines and take about 50 years for a one-way trip to Barnard's Star. The time for transmitting the data of the encounter to earth would be 6 to 9 years. The total project would take nearly 80 years from its start to reception of information from the star system. The starprobe would use 50,000 tons of fuel (deuterium and helium-3). The project proposal suggested that deuterium could be obtained from the earth's oceans, the helium-3 from the atmosphere of Jupiter.

At launch, *Daedalus* would depart from its parking orbit in the solar system and enter a solar orbit. Accelerating rapidly, it would reach solar-escape velocity in 2 days. The acceleration would continue for an additional 250 days to a distance of 0.0039 light years. Two of the first-stage fuel tanks would separate and tumble off into space, lightening the vehicle. Another tank drop would occur at 0.017 light years, and at 0.05 light years the first stage would have exhausted its fuel and been jettisoned into space. Separation of the first stage would take place 2.05 years into the mission.

Second-stage ignition would begin immediately after first-stage separation, and the burn would continue for 320 days. At 0.12 light years the first tank drop of the second stage would occur. The powered portion of the mission would end at 3.81 years at a distance of 0.2 light years. The final coast velocity would be 12.2 percent of the speed of light. The probe would coast for 45 years, taking measurements of interstellar space particles and fields and making size measurements of the galaxy using the increasing baseline of the probe from earth. The baseline would also allow measurements of stars' distances during the mission.

As the Barnard's Star system was approached, two large on-board space telescopes would begin a search for planets and determine their orbits. The larger, Jupiter-like planets would be spotted first. Earth-sized bodies would not be detected until *Daedalus* was relatively close to the star system.

Once the orbits had been successfully computed, a

series of small spacecraft would be launched. Using cross-range burns, *Daedalus* would position the probes for each planet in turn. In the final design, as many as 18 separate smaller probes could be carried by the main vehicle. All of the data would be relayed from the spacecraft to the main ship, where it would be stored for relay to earth.

Immediately following the encounter with the system of Barnard's Star, the main ship would transmit the data. After the end of data transmission, the main starprobe would continue to relay information on interstellar particles and investigations of the interstellar medium until it ran out of attitude-control fuel and began tumbling, or until it reached the end of its communication range.

Daedalus could, if it were programmed properly, execute a maneuver after the encounter with Barnard's Star that would allow it to perform a dual mission which would include an investigation of the 70 Ophiuchi system. The two stars are about 10 light-years apart, and the course change upon leaving the Barnard's Star system would be relatively small. Flight time at about 10 psol would take about 175 years.

After encountering 70 Ophiuchi it would continue onward for centuries, slowly disintegrating. In about 180,000 years it would exit the galaxy 13,500 light-years from the center in a direction in which little is known to exist within millions of light years, with the possible exception of dark bodies that might lie between the galaxies. Eventually the molecules of *Daedalus* would mingle with the elements of the universe, and all evidence of its manufacture by beings on the third planet of a yellow star out in the rim of the Milky Way would vanish.

The *Daedalus*-type starprobe is a highly sophisticated device. It is dependent upon a very expanded space civilization in the solar system; it requires the establishment of extensive manufacturing facilities in earth orbit and on fuel-processing stations in orbit near the moons of Jupiter. It will not be built in the next few decades, perhaps not within the next century.

If something like it is eventually built, it would be able to report back from Alpha Centauri in 35 years, mission time, including data transmission. It could send information home from several nearby stars within 100 years and from one of a dozen stars within 250 years.

A number of the nearby stars have potential as targets for dual missions such as the Barnard's/70 Ophiuchi combination. Some other combinations are: Luyten 726-8/Tau Ceti; Lalande 25372/+15°2620; Luyten 789-7/Ross 780; Sirius/Luyten 745-46; Epsilon Eridani/o Eridani, and others. If our future generations are rich enough, if space is developed enough, and if the technology is available, probes could be launched in a search that would be directed at the most likely candidates identified by advanced telescopes and arrays.

One far-future starprobe might involve a complex von Neumann device as described earlier. The starprobe would arrive in a target system and use the materials of that system to fashion a duplicate of itself. The duplicate would be provided with fuel and would be sent to another stellar system where it would duplicate *itself.* If each von Neumann device duplicated itself twice (by definition a von Neumann device can do anything, so there is little reason to suppose it couldn't make a dozen duplicates), the entire galaxy would be explored in times measured in a few hundred thousand years. The catch, of course, would be how to build even one von Neumann device.

THE FIRST STEP

"The first step, my son, which one makes in the world, is the one on which depends the rest of our days."

VOLTAIRE

If the progress into space proceeds in ways that have been envisioned, there will be manufacturing, lunar bases, colonies of workers, as well as power satellites to beam energy down to earth and spaceships which travel regularly to Mars and explore the asteroid belt for raw materials. When this level of development is available, mankind can, if it chooses, turn its collective mind to building the first manned starship.

A supply station on the moon would allow mankind to collect lunar resources and then package them for shipment by mass driver to other parts of the solar system—or to orbits near earth for construction bases. The mass driver, in modern space planning, has taken the place of the older, ubiquitous "space freighter."

Most people think of space as empty and useless—a medium to be survived, dangerous and inimical to mankind. It is far from empty, and it is not useless. Space is much like the oceans: while it is difficult for mankind to exist there without the support of a high technology, there are also tremendous potentials for energy and raw materials. The moon, the asteroids, and their minerals are a part of that potential.

Early in the planning for the conquest of space, it became clear that only 5 percent as much energy was needed to launch a payload from the moon as from the surface of the earth. Partly for this reason much time was devoted to talking about and planning a lunar base that could be turned into a mineral-processing center sometime in the future. If minerals could be processed on the moon, they could be sent to other orbits between the earth and the moon with a minimum expenditure of energy. In theory, the great expected industrial potential of space could be unlocked if raw materials from the moon could be utilized instead of having to ship everything needed up from the gravity well of earth.

The Apollo moon missions were, in part, designed to assess the availability of resources on the moon for future use. They did reasonably well, considering that 22 missions were originally planned and just 17 were flown, and that only on the last mission was there a crew member who was a professional geologist.

The regolith* of the moon has major constituents of aluminum, silicon, iron, titanium, magnesium, calcium, and oxygen. If an area of the earth were as rich in potential as some parts of the moon are, it would be regarded as an economically valuable deposit if the deposits we already use were not available or exhausted. The lunar soil is so fine that it does not need to be crushed, blasted, or ground for mining or mineral extraction. An extraction plant on the moon would also produce oxygen from the soil. Some estimates have put the amount of oxygen in the lunar rocks at 40 percent by weight over the entire surface.

Aluminum would come from the moon's anorthite, which contains 19 percent of the metal by weight. Bauxite, from which most aluminum is extracted, has 25 percent by weight. Because bauxite supplies are slowly decreasing, the mining and extraction of aluminum from anorthite has already begun on earth in several places. In Wyoming, Alcoa has begun mining a tract estimated to contain 30 billion tons of recoverable ore. Anorthite also has a relatively high percentage of silicon.

Titanium, very valuable on earth, is usually extracted from an ore called rutile. But it can also be extracted from ilmenite, which exists in quantity on the moon,

*Regolith is the layer of fine-grained debris that covers most of the lunar surface. It is the "soil" of the moon.

mainly among the basalts of the lunar seas. By weight, ilmenite is about one third titanium and one third iron. Techniques for processing titanium from ilmenite are within the realm of available technology.*

In the first enthusiasm over the moon's mineral "wealth" many people were perhaps too optimistic. The extraction processes proposed for the lunar soils would work, but it is a long way from a proposed chemical process to a plant on the surface of the moon complete with plant manager, foreman, and a working crew. And the lunar soils, while they do have substantial amounts of minerals, are not much more valuable than an equal cup of granite or clay on earth. The greatest virtue of the lunar minerals is that they exist on a body from which it is relatively easy to launch materials to earth orbits. The general scenario for those who have deep faith in the lunar resources begins with the space shuttle. The shuttle or its descendants would develop a small industry in the orbits around the earth. As power became available in quantity in space from solar power stations, the number of construction bases and space stations would increase. Lunar freighters would be constructed. As higher and higher orbits were inhabited by mankind, there would come a time when resources would be shipped there from the moon instead of from earth.

A lunar base and a processing plant of some kind has been standard science-fiction material for more than 50 years, and as early as 1962 there were concrete plans for a base that could not only begin mining minerals but also supply the earth with power from the sun. Studies by NASA also showed that lunar bases at the poles would be practical, and in the 1970s symposia on the use of lunar resources were held by the Lunar Science Institute and the American Astronautical Society.

*Carbon reduction of the iron, then chlorination, leaving titanium dioxide. The titanium dioxide is used with calcium metal at high pressure to produce titanium metal and calcium oxide. The calcium oxide is then leached away. All of the materials used in the process can be reused.

One of the early drawbacks to the plan for using the moon was the fuel needed for shipping the tons and tons of raw materials from the lunar surface. It may be much easier to launch from the moon (24 times easier), but trying to launch massive quantities of ore or heavy metal products is a staggering undertaking. Though there was some talk of generating hydrogen for fuel on the moon, possibly from polar sources of water ice, the problem seemed insurmountable.

Then, along came an idea that had been written about more than 30 years ago by Arthur C. Clarke. In the *Journal of the British Interplanetary Society,* Clarke described the basic mechanics of electromagnetic launch of mass from the lunar surface. He had apparently developed this idea from the postwar military research on electromagnetic launching of aircraft from carriers. The military research, in turn, was based on an understanding of electromagnetic fields that had been arrived at 50 years before Clarke began writing.

The Clarke proposal of a lunar "electromagnetic gun" became the "mass-driver." In concept it was something like a giant space conveyor belt, and would use magnetic impulses driven by electric energy to accelerate a small "bucket." The bucket would hold the mass-driver's payload of about 20 pounds of lunar material which would be ejected from the moon at the lunar-escape speed. Three hundred buckets in one mass-driver could deliver more than a million tons of lunar material to an orbit between the earth and the moon in only a few years. Over 10 years several mass-drivers could send enough material into space to build a complete habitat which might hold 10,000 people—a space colony.

Further studies of mass-drivers pointed out that they could be used as reaction engines to transport large cargoes, partly eliminating the need for a huge fleet of earth-to-moon freighters. A mass-driver could be attached to a cargo and off the cargo would go to its destination.

The mass-driver is the "great space hope" for the future. It can be used to move asteroids, launch materials from the moon and asteroidal materials to a waiting construction base. Here, a mining operation is taking place on an asteroid which has been moved to an orbit near the moon.

In 1977, a working test model of a mass-driver was built. It was 7 feet long with a drive-coil diameter of only a few inches. It could accelerate a 1-pound bucket mass at 35 g in 0.11 seconds. Although this was a laboratory model, it suggested that the idea might be practical in the future.

The advocates of the use of lunar materials to build space colonies also developed the idea of a "mass-catcher." One model was a huge, self-propelled craft 300 feet wide and a quarter mile long. In pairs, mass-catchers of this type could be used to recover the lunar materials at a specific point between the earth and the moon. The space-colony advocates favored a point called L-2, 40,000 miles behind the moon.

The mass-driver payloads would come in at speeds of 600 mph, be captured by the mass-catcher, and when it was full (one design resembles a dirigible with one open end), it would depart for an orbit where the minerals from the moon would be used for construction. The catchers have been described as "supertankers of space."

At least in theory, the raw materials were there. There was a prototype technology that might develop into a means of using the materials; there was the mass-driver for reaction engines and for slinging ore around the void.

Not long after it was realized that lunar raw materials could be refined and used for building space structures—including spaceships—another long-dormant idea experienced a resurrection: asteroid mining. Estimates showed that raw materials could be as easily retrieved in some cases from the asteroids as from the moon. The asteroids had other possibilities as well; in theory, they could be hollowed out and turned into "off-the-shelf" space stations.

Between the orbits of Mars and Jupiter is a vast ring-shaped gap 340 million miles wide. It is filled with hundreds of thousands of chunks of rock called the asteroids—sometimes the planetoids or minor planets—which are left over from the information of the solar system. Most of them stay within the boundaries of the gap and are indistinguishable from one another except by size. The general group orbits the sun in 1½ to 6 years. One oddball takes 13 years, and a few take one year or less. One comes closer to the sun than Mercury; one has an orbit which takes it out to Saturn's distance.

The first asteroids discovered are relatively large, and at least two of them are probably spherical, truly minor planets. Ceres' diameter is estimated at about 500 miles, Pallas' at 300, Vesta's at 250, and Juno's at slightly over 100. The great majority of the bodies are very small, irregularly shaped objects. It has been estimated that 70,000 asteroids are within reach of the earth's telescopes. Of these, fewer than 50 are as large as 25 miles in diameter. Only 3000 asteroids have been sufficiently well observed with telescopes that an accurate orbit is known.

Two groups of asteroids are particularly interesting because they can come close to the earth and are relatively easy to reach by manned spacecraft. These are the Apollo and Amor groups of close-approach asteroids. Apollo objects have orbits which intersect the earth's at some point; Amor asteroids have orbits that cover some portion of the path between earth and Mars.

Since 1932, a total of 28 earth-crossing Apollo objects and a smaller number of Amor bodies have been found. Between 1973 and 1976, a serious program was launched to determine the characteristics of these. During the search, 12 new objects were found, which included two new earth-crossers, 1973NA and 1976AA. The latter was the first discovered Apollo object that did not cross the orbits of Mars or Venus as well as that of earth. Over the years during which Apollo asteroids have been found, scientists have estimated the total number, and the estimates vary from 50 to 750 ± 300.

The asteroids in general were ignored by astronomers

for over a century. Investigating them was, so the remark went, "like rummaging . . . in the cosmic garbage." Gradually, however, it was recognized that the asteroids might help explain planetary formation. The Apollo objects particularly had an importance out of proportion to their size and number because it was this type of asteroid which had, in the dim past, produced the numerous craters—at least those over 3 miles in diameter—on Mercury, Venus, the moon, and earth. The surface density of the impact craters was a direct means for determing the age of geologic regions. Crater counts could help us compare the geology and evolution of one planet with another.

The other interesting aspect of the Apollo bodies was their relative accessibility by spacecraft. Some of the objects had orbits so similar to earth's that they were easier, or at least as easy, to reach as the moon. Since they were so close, there was speculation about their composition and what a manned spacecraft might find if it approached one.

When a dozen of the Apollo asteroids were subjected to observation that gave clues to their composition, it was found that they were similar to ordinary asteroids. Remote-sensing evidence from meteor trails also gave a clue to the materials of Apollo bodies, as did meteorites (believed to be chunks from Apollo asteroids) and analysis of high-altitude meteoric dust. Assuming the Apollo bodies were composed of materials similar to meteorites and asteroids in the main belt, scientists felt the nearby bodies should be a source of water, carbon, carbon compounds, and some free metal: nickel-iron. The minerals of major significance would include: olivine (iron-magnesium silicate), pyroxene (iron-magnesium-calcium silicate), feldspar (calcium and aluminosilicate), clay minerals, and various other metals in small quantities.

From their brightness it was estimated that most of the known objects were ½ to 1 mile in diameter. The diameter of an asteroid can be estimated from its brightness if it is too small to measure directly in the telescope. An object composed, for example, of typical lunar material would show a certain brightness from its distance only if its diameter were fairly close to a specific value.

The composition of the asteroids of the solar system led to proposals for mining them for raw materials. The main belt may have more than 400,000 bodies larger than a ½ mile in diameter—an immense resource for a space-based civilization. Not only could the Apollo bodies be reached fairly easily by spacecraft but the Apollo bodies were close enough so it would be conceivable to divert their orbits so that they would come near enough to earth for mining.

A very early and crude method, using hydrogen bombs to bring asteroids into earth orbit, was proposed. A fairly high-yield megaton bomb would be detonated on the asteroid's surface. The resulting crater could become a temporary thrust chamber for the detonation of other bombs. Even a large body like Eros could be "captured" by using several thousand megatons. The method had several drawbacks: the nuclear-test-ban treaties would be violated; no one had any experience in using nuclear bombs for propulsion of a solid body (although in theory it should move); and there was a danger of setting off an explosion along a natural fault line in the asteroid and winding up with a large number of dispersed meteors all departing in different orbits.

As early as 1967, a study considered a trip to the small asteroid Icarus, principally known as a faint streak on a 1-hour star-field exposure taken with the 48-inch Schmidt camera at Palomar Observatory. Icarus passed closer to the sun than any other astronomical body, was an earth-crosser, and in 1968 would pass within 4 million miles of our planet.*

*At its closest approach to the sun, Icarus is twice as close as Mercury.

A mass driver (on the left) can be constructed using materials from earth brought by a space shuttle. A solar-powered mass-driver (in the center) is heading for the outer reaches of the solar system for an asteroid intercept.

The study concluded that it would be possible to reach Icarus with the then-existing *Saturn V* launch vehicle and the Apollo spacecraft *CSM*. Although it would not be easy to use a small, three-man spacecraft like the *CSM* for an asteroid mission, the conclusions demonstrated to those interested in the development of space that it was possible at least to think of asteroid missions and perhaps retrieval.

Modern ideas of asteroid mining usually involve the use of a mass-driver to send ore to earth. The mass-driver would use materials from the asteroid for fuel. Landing on an asteroid a few miles in diameter or smaller is much easier and safer than landing on the moon. There is practically no gravity to fight, and landing is more a matter of matching speeds and gently drifting closer to the body. The needed velocity change (the ubiquitous "delta-V," in NASA talk) from an earth orbit to that of an Apollo asteroid is small and consequently within the propulsion capability of current technology—certainly within the technology of an advanced space industry.

Once landed on an asteroid, the crew of a manned mission would prepare it for delivery to earth orbit. Low-delta-V bodies currently known are: Adonis, Eros, Hermes, Apollo, Geographos, Toro, Amor, Ivar, and the numbered bodies 1976UA, 1977HB, 1959LM, 1973EC, and 6743PL.

An asteroid retrieval would proceed as follows: the parts of an asteroid-retriever/mass-driver with fuel for the outbound trip and materials-processing equipment would be launched into low earth orbit (usually called LEO) by the space shuttle or a descendant vehicle. When the package was assembled in LEO, the asteroid-retrieval crew would be brought up by shuttle. The retriever would leave the LEO with a delta-V of 6.4 km/sec and begin a slow spiral orbit up to a high earth orbit. This trip would take about 2 weeks using a mass-driver.

The retriever would use a gravity assist from the moon for escape velocity to the asteroid. A delta-V of 3 km/sec would bring the retriever to the target. Additional gravity-assist techniques using both Earth and Venus would be employed during the outbound leg of the trip. The mission would achieve orbital matching in about a day, with a small velocity change.

Once the rendezvous with the asteroid was completed, the crew would land at a location near the pole of the body. A conventional explosive would be placed to break off a fragment near the equator weighing about a million metric tons. The crew would then use a small vehicle to land on the detached mass and "despin" it (asteroids rotate, and the fragment would still have the rotation). Using cable, they would wind up the fragment and then

The keys to man's expanding space civilization which may one day build a starship. In the foreground is an asteroid being mined for raw materials. Above it is a giant solar power station providing energy for the operations. The station is about 7 miles long by $2\frac{1}{2}$ miles wide.

unwind it. The process of despinning would take about 6 days. The fragment would be put in an aluminum-cable net and the mass-driver would be attached. The mission would depart the asteroid's orbit, then use planetary gravity assist and lunar gravity assist to return the fragment to a location near the earth where a space-manufacturing facility had been established.

As a hypothetical example for the asteroid 1977HB, an asteroid retriever would depart LEO April 28, 1984, and would return the 1 million metric tons of raw materials to near earth by mass-driver by September 23, 1989.

Since all of this is within reach of space technology, either now or in the foreseeable future, the idea of masses of raw materials brought to the earth from space or from the moon is a relatively practical one. And if massive quantities of structural materials can be made available in space, then some of those materials could, in the future, be used to build a starship.

RENDEZVOUS WITH A STARSHIP

"All this world is heavy with the promise of greater things, and a day will come, one day in the unending succession of days, when beings, beings who are now latent in our thoughts and hidden in our loins, shall stand upon this earth as one stands upon a footstool and laugh and reach out their hands amidst the stars."

H.G. WELLS, 1903

The basic idea of using raw materials available in space to further the development of space is not new. Robert Goddard proposed using extraterrestrial resources to manufacture propellants and structures as early as 1920.

But following the lunar landing in 1969, and the development of the space shuttle, much more attention was paid to notions which might have seemed far-fetched earlier.

The idea has always been tied to space habitats—large structures in orbit housing hundreds, perhaps thousands of people. The literature, design studies, and investigations into large space habitats are important for starships. If a successful, closed-life habitat for a large number of people can be developed, then it can make a voyage to the stars when a propulsion unit of sufficient thrust is attached. A spaceship, even a tiny Apollo capsule, is essentially nothing more than a closed habitat with engines.

Discussions about living in space predate those of starflight. Jules Verne wrote about self-contained worlds in space in 1878, as did Kurd Lasswitz in 1897. The first mention of a space *habitat*, however, was made by Edward Everett Hale in 1869 in a short story called "The Brick Moon." These three literary efforts were not intended as scientific proposals, but Tsiolkovsky followed them in 1903 with a discussion which covered the basic principles of any permanent habitat in orbit: rotation for artificial gravity, a closed ecological system with a greenhouse for food production, and the use of solar energy for power.

Once again in the history of space exploration, Goddard followed Tsiolkovsky. Goddard envisioned a great space habitat which could be powered toward the stars on a generations-long voyage. His "ultimate migration" theme was repeated by J.D. Bernal whose name survives today in a space-colony design called the "Bernal sphere". In a book oddly entitled *The World, the Flesh, and the Devil*, he suggested a hollowed-out asteroid could be used as a starship; Bernal made his proposal in 1929.

During the 1920s, '30s, and '40s, various scientists kept returning to the concept of a permanent home in earth orbit. German rocket pioneer Hermann Oberth proposed small scientific research stations in low earth orbit. An Austrian named Potočnik, writing under the pen name Hermann Noordung, proposed what later became the familiar wheel-shaped space station made famous in the movie *2001*; Wernher von Braun popularized the wheel-shaped space station in the U.S. in articles in the old *Colliers* magazine and in a book, *Across the Space Frontier.* The U.S. Air Force funded a small manned space station called MOL, for "manned orbiting laboratory," until it was eliminated from the budget along with the vehicle which would have supplied it, the Dyna-soar (an early version of the present-day space shuttle in concept but not in looks). Skylab, of course, was a true space station and was a permanent home for astronauts for several long-duration missions in the early 1970s.

All of these proposals, designs, and the one actual station launched into low earth orbit had one thing in common: they were designed to be built on earth from materials here and then launched at great energy expenditure. Ideas for permanent stations after the space shuttle was operational were still confined to the same premise. In addition to the obvious factor of cost, earth-launched materials would require numerous trips by a shuttle-type vehicle, and there is a limit to the size of habitat which could realistically be put in orbit. None of the sizes imagined could hold enough of humanity to be a multigeneration starship. Even if automatic beam builders and other space construction devices became available during the design life of the space shuttle, the materials for constructing a large station must still come up from earth.

In 1956, Darrell Romick proposed that a large colony of humanity could exist in space aboard a giant cylinder more than half a mile long and almost a thousand feet in diameter; the habitat could hold 10,000 people. Architect Paolo Soleri proposed a great space city which he called Astromo. It could have a population, Soleri estimated, of 70,000 and would be 8000 feet long with an internal surface area of 466 acres. Both of these projects were much different from the low-earth-orbit space stations. They could certainly not be built in space of materials shipped up from earth. For a time, they remained farfetched dreams.

About the time of the lunar landings, Gerard O'Neill was teaching physics at Princeton. He posed the following question to his students: "Is the surface of a planet the right place for an expanding technological civilization?" His students, somewhat to his surprise, produced the answer that inside-out planets might be more practical. O'Neill began thinking about artificial worlds in space during the period when the surface of the moon was being investigated by Apollo astronauts. The result of his work was the so-called "space colony" built of lunar materials transported to earth orbit by mass-driver.

O'Neill's first proposal, called "Island One," would be populated by 10,000 people divided into three villages; the whole colony would be self-supporting, once established. Sunlight would come into the colony from planar mirrors shining through large outside windows set in the upper latitudes of the spherical world.

His later colonies, especially "Island Three," were even more ambitious: paired cylinders 20 miles long and 4 miles in diameter. The total land area would be 500 square miles—about half of which would be living space for a colony of 4 million. Some scientists studying O'Neill's proposals thought that "Island Three" could conceivably feed, though not necessarily maintain, a population of 10 million. Food could be exported from the colony down to the starving billions of earth.

The attention given to O'Neill's ideas of space colonies sparked intensive research into large space habitats and how they could be constructed and supported independent of earth. Never before had the design of habitats been stretched to the scale of a gigantic colony housing thousands of people. In 1975, NASA and the American Society for Engineering Education sponsored a program in engineering systems design at Stanford. The goal of the study group was to design a plan for the colonization of space. The results were comprehensive, persuasive, and thorough. The conclusion was that large space habitats were not only possible but could conceivably be built with technology available now and through some reasonable future advances. The space colony was, in fact, a twenty-first century idea which might even be accomplished within the next 50 to 100 years.

Following the NASA design study, other scientists and research groups added their ideas. From a hundred or more different and very bright people came a large body of concrete suggestions, basically practical and within current technology, for making a hollow object in space livable for a large population. The relevance for starflight is obvious; it makes little difference whether the hollow object is a space colony or a multigeneration starship.

Mankind now had the basic ingredients for starflight. There were proposals for retrieving asteroids and proposals for making large hollow bodies livable for long periods. The lunar landings had resurrected the idea of using the moon as a source of raw materials for a space-based industry, and there were suggestions that fusion energy might be a near-future possibility for propulsion. A few studies, as we have seen, also showed that at least one other star nearby might have planets, and man was beginning to understand how to predict the possibilities for habitable worlds through his successful studies of the planets of the solar system.

One of the first people to revive Bernal's old interstellar-ark ideas after the Age of Space had begun was Dandridge Cole. (L. R. Shepard of the BIS proposed a Noah's-ark nuclear-powered interstellar colony in 1952, but he was ahead of his time and his proposal was based more on imagination than on engineering; like Bernal, Shepard envisioned a hollowed-out asteroid as the "starship.") Cole was also the first to comment on the economics of building a starship. He discussed asteroid mining, raw-materials processing in space, and using hollowed-out planetoids as space colonies in earth orbit. In a 1964 book, *Islands in Space*, Dandridge Cole and coauthor Donald W. Cox discussed many of the features that would later appear so often in space-colony literature: linear electric motors (mass-drivers); space catapults; artificial gravity in a cylindrical, spinning world; and others. He also discussed using a colony aboard a hollowed asteroid to travel to the stars.

Carving out the interior of an asteroid for a space colony or starship has been compared to carving ocean liners out of granite blocks; the analogy is perhaps a little unfair. Though no modern ocean-going ship has ever been carved out of a single unit, the earliest ships of mankind were made from single logs—dugouts. It is not inconceivable that our first starship also could be constructed from a solid object.

One of Cole's more typical colonies was 20 miles long and 10 miles in diameter. Solar energy was collected by a large mirror at one end and reflected down the horizontal axis to provide sunlight for farms and colonists. The interior was much like those pictured in books about space colonies; the curved horizon of an inside-out world, fields and flowers, trees and small cities. Artists from both General Electric and the Martin Company (now Martin Marietta) produced illustrations of the exteriors and interiors of asteroid space colonies in the late 1960s.

The process of hollowing out a minor planet has some interesting engineering possibilities. John Campbell once suggested that large mirrors could be produced in space that could concentrate the sun's rays for cutting and shaping the body. With automatic beam builders already available or being designed for use with the space shuttle, it is possible to imagine that a giant solar mirror could be constructed in high earth orbit in the future and that it could be used at a construction orbit to work on an asteroid.

The method advanced by Dandridge Cole also used giant mirrors. The supports of the mirrors would be formed in space under zero-g conditions so that they could be made from light materials, because there would be no gravity pulling them out of shape. The mirror material could be one of the extremely thin, lightweight, silvered plastics, such as Mylar. The mirrors would bore a hole down the center of an asteroid 2 miles long and a mile in diameter. When the central bore hole was completed, the cavity would be filled with tanks of water—perhaps water produced from rocks of other nearby asteroids. Once the tanks were in place, the body would be set spinning by conventional propulsion devices using hydrogen as fuel (the fuel might also come from materials nearby). Spinning in the light bath of the solar mirrors, the asteroid would heat up, so that eventually the surface would turn a dull red, then bright. The melting point would be reached and the intense heat would start inward, eventually reaching down to near the core.

With the egg-shaped asteroid nearly completely molten, gravitational and cohesive forces would tend to pull it into spherical shape; however, the central core of water would be the last part to be heated. The net result would be a tendency for the molten asteroid to maintain its cylindrical form. If the timing were perfectly correct, the tanks would explode from steam pressure just as the central axis of the asteroid melted.

If this little bit of tricky timing could be accomplished, the product would be an asteroid expanding like a balloon, ending up with a fairly thin "hull" or crust and a diameter of about 10 miles, a length of 20. Using an iron asteroid, Dandridge Cole's theoretical process would open a complete space colony, waiting for the crew to go into the cavernous interior and begin construction.

For artificial gravity, the colony would spin at 9 revolutions per hour resulting in a "day" of 6.6 minutes. The surface area formed from the blown-up one-mile-by-two asteroid would be 628 square miles, over half the size of Rhode Island. By Cole's estimate, the colony could hold 100,000 people. The mirror that had produced the colony would be transferred to one end of the asteroid, where it would shine a beam down the long axis of the colony to make a linear sun.

There are certain attractive qualities about a hollowed-out body—the hull is "off the shelf"—but there are also severe drawbacks. The interior hull of the object would have to be coated to prevent leakage through natural fissures and cracks in the surface. Hollowing out something as large as an asteroid, though an interesting concept on paper, is a hard process to put into effect. And when it is all done, and the energy and time have been expended, the result would probably be as expensive and difficult as constructing a hull from refined raw materials. Apollo asteroids and others may serve better as raw materials for construction than as starships, although the idea is still an attractive one to some.

The most likely prospect for a starship is one that a space-based industry and civilization would build after a large and complex development of space took place in the future and space colonies, or permanently manned large habitats, had already been constructed. The experience gained in building space colonies would provide the working technology with which to build a habitat that could travel beyond the solar system.

How soon this might happen depends less on technology than on a commitment to exploring and using space. Estimates in NASA studies indicate a low-earth-orbit station housing 200 people could be built in 20 years. A lunar base that would contain an initial crew of 300 and an operating crew of 150 could be built over a period of 10 to 20 more years. If the low-earth-orbit station and a lunar base were followed by a space construction base

When mankind is able to build and maintain large structures in space and produce raw materials and finished products in high orbits and other locations in the solar system, a base where a starship like the Agamemnon can be constructed will be possible.

housing 2000 people, built up over another 10 years, then a space colony could be the next step. A colony housing about 10,000 people could be in orbit near the moon in perhaps 75 years.

Several scientists interested in space colonies have investigated the prospect of using the construction base to build solar power satellites that would provide an energy-starved earth with electricity. If this were done, then a colony would be built more slowly, based on a partially self-supporting economy. Some studies by NASA and others have indicated that eventually a large colony in space could be entirely self-supporting; it could be the initial component in a future space-based civilization that could build structures in earth and lunar orbit, spaceships to explore the planets, and other colonies for specific purposes.

Colonies devoted only to scientific investigation, colonies to direct energy production and distribution to the earth, and colonies involved only with bringing in raw materials for manufacturing or asteroids for mining could be established. Eventually a large habitat could be constructed for the purpose of building a starship. It would draw asteroids into earth orbits for mining; the materials of the asteroids would be processed, and some would be sold to other colonies or to various space-based industries. In this highly developed space-industrial complex, a portion of the profits from mining and processing operations and from sale of minerals and finished products could be diverted to the building of man's first attempt to reach the stars.

The prospect of such an effort staggers the imagination. The starship construction base would require 10 million tons of shielding, 220,000 tons of soil, and half a million tons of structure. A population of thousands would have to be transported from other colonies to the starship orbit. Surrounded by construction bases, the starship would slowly take form as one of the most massive projects imaginable—the end and the beginning of mankind's great dream of travel to the stars.

As a habitat for human beings the starship must hold an atmosphere; this effectively limits the shapes that it can assume. To economize on structural mass, a large shell holding a gas inside under pressure must act as a membrane in pure tension. There is a direct relationship between internal loading and the shape of the surface curve of a membrane configuration. There would also be effects upon the shell caused by a spin to simulate gravity. Four shapes are acceptable: a sphere, a cylinder, a torus, and a dumbbell. For various technical reasons, a dumbbell is one of the least attractive. A torus, often seen as the projected design for a space colony, has some disadvantages for a starship, but a cylinder has the smallest forward-facing area in the direction of motion. For this reason, it is easier to shield from interstellar dust, particles, meteoroids, and motion-produced effects in the interstellar medium.

There have been several serious studies dealing with large habitat fabrication. Some of the ideas involve extrapolating current manufacturing trends; some depend upon methods yet to be developed but probably within the scope of a twenty-first-century space-based industry. The extrapolation of present manufacturing involves the use of aluminum ribs and plates for the hull. A future alternative might be a hull made using metal-vapor molecular beams.

In the metal-vapor process the hull of the starship would be a seamless stressed-skin. The system would consist of a large solar furnace that would provide heat to an evaporation gun. The evaporation gun would direct a conical molecular beam at a mold—perhaps a giant Mylar balloonlike form in the shape of the ship's hull. The balloon would be rotated under the molecular beam as a metal plate on the surface gradually built up to the required thickness and strength. If the hull were a deposit of aluminum, the balloon form would need to be at earth room temperature for the best quality deposit. The difficulties in vacuum-metal fabrication now are a lack of uniformity and getting a good strength to the deposited metal film. Nothing of great size has been attempted using the process.

In a more common form of construction, aluminum ribs and plates would be made in space factories from the raw materials of either asteroids or the moon. The starship might look much like one of the giant external tanks of the space shuttle: an end cap followed by a succession of aluminum rings and another end cap.

The forward end cap would be constructed first; then a structural ring would be attached. A large factory would simultaneously roll aluminum plates into a ring and join them. Once one ring was attached to the forward cap, the open face of the ship could be closed off to make a habitable working environment for construction of the interior. The ship would grow as the rings were added, rather than having the hull completed in one process, as would be the case if metal vaporization were used for the construction.

Once the hull of the starship was completed, the cylinder would be spun to simulate gravity. It would not be necessary to simulate full earth gravity aboard the starship; and there are certain circumstances under which it might be desirable to have only two thirds of Earth gravity on the inner surface of the cylinder. It is unlikely that a starship would travel in zero g. What happens to the human physiology in the absence of gravity is not well understood, but our limited experience with spaceflight indicates that decalcification occurs in humans living in zero g at a rate of 1 or 2 percent per month; this results, of course, in decreased bone mass and density and weakness in the skeletal structure.

The absence of gravity brings other physiological problems: hormone imbalances occur; protein and carbohydrate states are unstable; and there are signs of low blood sugar.

Rotating the starship for artificial gravity poses some problems. In a rotating system a person would feel a sensation of weight much like gravity; when the person moved, another force would be felt—the "Coriolis Force."* If the spin were rapid enough, simple movements would become a complex series of motions. The Coriolis force can also affect the fluids of the inner ear and cause confusion. Turning the head can make stationary objects seem to gyrate wildly. Studies of the effects of rotation on medical subjects on earth do not give a clear indication of what might happen in space over a period of time, but they have indicated that a rotation speed of 1 rpm or less would be the best upper limit for the areas of the starship in which the crew would spend most of their time—particularly the residential areas.

With a spin of 1 rpm to simulate gravity, a cylinder with spherical end caps about 3000 feet in diameter and several miles long would have an immense interior and could support a colony of more than 200,000 people with plenty of freedom inside. For the initial crew, it would represent a vast environment compared to the average amount of space per person available in a city in most countries of the world.

To maintain life aboard the starship the atmosphere must sustain the crew without long-term physical damage and must be renewable. Earth's atmosphere is 78 percent nitrogen and 21 percent oxygen. Since the conditions aboard the starship should be as earth-like as possible for human longevity—the earth's systems work fairly well as a "spaceship" and have for millions of years—the atmosphere will be similar to the one we

*Coriolis force can have very strange effects aboard a rotating cylinder like the starship. If a crew member were standing on the inner wall of the ship and tried to shoot an arrow in the direction of the central axis, the arrow would deflect from a straight line. It would have a curving trajectory bent in the direction of the rotation.

enjoy now. Its outstanding features would be partial oxygen pressure, as compared with earth's, a slightly higher partial pressure of carbon dioxide (to help plants in the starship farm), and a partial pressure of nitrogen about half that of ours at sea level.

There are other possible combinations for the atmosphere, however, including nitrogen-oxygen in different proportions and a mixed-gas similar to that used by deep-sea divers (helium-oxygen mix), but the physiological factors and effects of different atmospheres on a long-term basis are not known. It is better for mankind, perhaps, to stay with the hospitable natural environment developed on earth over millions of years.

The ship's atmosphere must contain some of the minor gases and water vapor. Without the presence of a small percentage of water vapor, the membranes of the lungs draw moisture from the body and lose it in exhaled breath. The result is constant dehydration unless large amounts of liquids are consumed. On earth, water vapor normally arrives in the atmosphere from the oceans and returns as rain. Aboard the starship, the water vapor could also come from bodies of water on the interior surface.

Both the carbon dioxide balance and the heat balance of the ship must be rigidly controlled. Carbon dioxide on earth is absorbed by plants which produce oxygen by photosynthesis. It is produced as waste by human breathing. The average production of carbon dioxide by a human being in one day is 2.2 pounds, whereas the oxygen intake is somewhat less. The result in a closed human ecology is an excess of carbon dioxide. Plant growth would have to be carefully maintained to prevent carbon dioxide buildup. Chemical reduction of excess carbon dioxide to oxygen is possible, but the process is much more difficult than tending plants carefully.

Both water vapor and carbon dioxide contribute to heat in a closed environment. The water vapor absorbs heat radiated from the interior of the ship and can affect the heat balance. Carbon dioxide in quantity also tends to absorb energy and can have an effect on the heat balance.

A minor but plaguing difficulty with the starship atmosphere can come from particulates—suspended liquids and solids also called "aerosols." One of their effects is to serve as condensation points for fog, and some can be a health hazard. On spaceship earth, particulates are removed by precipitation, but precipitation may not be desirable aboard the starship. The earth's gravity takes care of larger particulates, leaving them on the ground instead of in suspension in the atmosphere. Aboard the ship, the gravity, though present as a simulated effect due to spin, will decrease toward the central axis, at the "center of the atmosphere." The net result will be sedimentation by pseudogravity near the outward surface, but not in the "upper air." Particulates in the air could be removed by electrostatic devices similar to those now used to clean the air aboard nuclear submarines.

The circulation of the atmosphere is relatively critical. The valley areas on the surface (the inside hull walls) where pseudogravity is the highest will absorb more sunlight than other areas, assuming that "sunlight" is introduced somewhere along the central axis of the ship. The air will rise over the warm areas, travel through the hollow interior space, and cool. It will sink over other parts of the interior surface.

The light in the ship could radiate from high above the central axis through a transparent and hollow cylinder. The cylinder would be filled with a gas of the proper density to produce efficient scattering of the light in a duplication of the earth's own atmosphere. The poles of the interior space where the gravity is zero would be shaded and equipped with thermal radiators mounted on the exterior hull to balance the complex heat exchange that will go on in the interior.

With the shaded poles, water vapor in the starship's atmosphere would condense and probably freeze there. When the mass of snow and ice at the poles grew large enough, it would drift "down" into the interior of the ship, and sunlight on it would turn it to water again. The water would then run to the rivers and lakes of the interior. Clouds could form, but mostly over the warmer areas of the interior. Cold clouds on earth are made up of ice crystals, and it is unlikely the interior temperature would be low enough to produce ice crystals.

The starship would need a carefully controlled food-processing environment to survive a long voyage; it must also have a complex system for recycling waste. A space colony can replace raw materials with new supplies brought from earth or from another colony. Traveling between the stars, the ship will have little chance of replenishing anything lost.

Research into life-support systems in space has been going on much longer than is generally realized. As early as 1912, Tsiolkovsky published a review of some of the requirements. He anticipated many of the ideas that went into the design of the Apollo spacecraft and foresaw the eventual need for regenerative and recycling support systems for long-term living in space. He considered the use of biological devices to produce food and recycle wastes into usable materials.

Partly because of the early impetus given by Tsiolkovsky, the USSR has investigated the life-support problem for somewhat longer than has the U.S. Some of the first studies involved an algae/higher plant/man-involved system. Chlorella-sp algae were used to recycle organic wastes and revitalize the atmosphere. The higher plants, including vegetables and wheat, produced food and recycled nutrients and air. But algae are not outstandingly productive plants, and humans are fond of what they are used to eating. The best choice for an enclosed environment is probably an advanced terrestrial agriculture based on plants and meat-bearing animals.

One presently existing earth agricultural technology which is applicable to starships is called Controlled Environment Agriculture (CEA), and has been researched in this country and Western Europe. The main thrust of the research is food production in a microclimate inside a dome or other sealed system. Humidity, water vapor, wind, circulation, temperature, and carbon dioxide concentration are all controlled. The yield in such systems compared to that of "normal," open-field agriculture is amazing. In open-field agriculture it is usual to harvest 1 crop per year. With vegetable crops under the CEA system, 3 crops a year can be expected, and the yield can be 6 times greater to almost 20 times greater depending on the crop (cabbage, 6 times as much; cucumbers, 20 times).

Most countries have experimented in recent years with various types of "super farms," including the CEA system, to increase food production for the coming century of expected population growth. Traditional Kansas farming can produce slightly less than 2 tons of wheat per acre; corn, 3½ tons per acre. Improvements by such groups as the International Rice Institute in the Philippines have resulted in yields averaging more than 16 tons per acre, and the research, in some ways, has only just begun.

Using multiple cropping, a sequence might start with rice, followed by sweet potatoes, then soybeans. Corn would be next, and then soybeans again. The yield might be 2 tons per acre of rice, 9 tons of sweet potatoes, 4 tons of soybeans in two plantings, and 20,000 ears of corn.

Some of the reports of increased yield through the use of artificial lighting 24 hours a day, rapid and high-volume ventilation, and controlled termperatures are incredible. In laboratory experiments in England, yields as high as 1000 pounds per acre per day have been reported using an atmosphere with a carbon dioxide content of 0.13 percent. A high of 15,400 pounds per acre per day has been produced in hydroponic gardens at Arizona State University.

Not all of the research into crop yields and vegetable modification is done at universities where the chemical nutrients and lighting are available. In Mexico, there is a "wonder farmer" living in Valle de Santiago who has produced 60-pound cabbages, 10-pound onions, and collard greens the size of palm fronds. J. Carmen Garcia became something of a celebrity in 1980 when it was discovered by the international press that he was able to produce giant vegetables. Reports coming out of Guanajuato, 260 miles northwest of Mexico City, told of Garcia's competing on identical plots of ground, with the Mexican Ministry of Agriculture. The contest was held at the Tangasneque experimental station near Tampico.

The result of the Garcia "grow-off" with the Mexican government showed he could produce 107 tons of cabbage per hectare (2.5 acres), whereas the university-educated farmers from the Ministry of Agriculture could produce only 5 tons per hectare. Garcia claimed he had a special formula in which the seeds were "energized" before planting. Fighting a cavalry charge of press photographers and reporters, the superfarmer remarked that the secret, if there was one, "was for the benefit of humanity."

The critical part of designing a closed environment for the starship crew is human diet. Once a general diet is determined, it is possible to design the closed system. Developments in the field of agriculture have been so phenomenal that only 100 acres could basically support 8000 to 10,000 people. Thus, the choice involves diet and available acreage. The starship can utilize three-dimensional food production from plants, unlike earth, and the "farm" could be relatively small in overall area. Vertical agriculture would utilize parallel rows of vertical cylinders. The idea of vertical agriculture has been under research for several years to improve food production in extremely mountainous communities with high population density.

One of the vertical-growth agricultural systems already used is called the "purulator system" (meaning to grow upwards profusely). Vertical growth can yield an increase of 10 to 20 times more than horizontal-intensive farming on earth, which has already brought impressive increases over traditional horizontal farming. A vertical system can produce on a year-round basis, and the cylinders allow for quick removal of a farm column in the event of crop disease.

There are other possibilities: we may be able to breed fish without water in the weightless areas of the ship. Gravity on earth causes the collapse of gills when a fish is removed from water. The gills would not collapse in the absence of gravity, and they could be bred in the parts of the ship's atmosphere where the humidity is high. It has been impossible, for obvious reasons, to test "hydroponic fish," but it can be done as soon as a space station is regularly manned above the earth.

The diet aboard the starship could be varied extensively, although meat products would be a luxury. Protein production in animals is not particularly efficient. Generally it takes 10 pounds of feed to produce 5 pounds of live weight in cattle. The yield of edible meat is about 49 percent, so it takes 10 pounds of feed to produce 2 pounds of beef. Lamb produces one of the lowest yields of edible meat per feed intake.

Recycling of wastes from plants and animals is a complex process, but the technology is not a distant one, except in the size of the application to a space colony or starship. Water can be reclaimed from the air by dehumidifiers in the farm areas. Most of the water in the atmosphere will come from the process of transpiration from leaves of plants. Water can also be extracted from wastes of all kinds by the "Zimmerman process." The end result of the procedure is a relatively pure water with some ammonia and phosphate ash and a gas with a high percentage of carbon dioxide. The gas can be used in the starship farms and the water by the crew.

The Zimmerman process does not necessarily require energy, either. If the waste fed into the recycling system is high in combustible solids, the process is self-sustaining without an outside source of heat; the heat of combustion will keep the system going.

With the hull finished, the atmosphere in place, and some of the interior completed, the starship will become semioperational. Its monitoring and control subsystems will be very similar to the nervous system of the human body. In many ways it will be a living spaceship and will have to be treated as one. With the exception of the hull and engines, it will be an organic machine and one whose performance must continue year after year.

Mankind's first starship will be the product of a myriad of different major advances in thinking and technology that have happened in this century, principally in the last two decades, but it will also depend on the development of technologies and designs still in the future. Most of all, the starship will be a product of the dreams, discoveries, and dedication of hundreds of brilliant scientists and inventors over the last two hundred years and the thinkers, technicians, and dreamers who will follow them.

From pieces of an asteroid and the remains of lunar materials, mankind will have built a great hull out in an orbit around the sun. The hull will be filled with crushed lunar and asteroidal soil, moistened with water taken from the raw rocks and surfaces of other bodies. The great cylinder will almost be ready for an interim crew. It will become a symbol of mankind, perhaps a once-only project. None of those working on it, or in it, or completing the interior will ever labor at anything so distinctively representative of human qualities of adventure, inquisitiveness, and courage. And when it is finished it will have a name. It will be called the *Agamemnon.*

A STATELY PLEASURE DOME

"Through the clear air you can easily see the pole cap about five miles to the north and the equatorial sea almost an equal distance to the south."
DANDRIDGE COLE, *Islands in Space* (1964)

The interior of the *Agamemnon* after a decade will be an inside-out world, with an up-curving and encircling horizon. Objects, including residences on the "opposite" side of the world, will hover overhead in the distance. The "polar caps" will be visible, and an equatorial sea will completely encircle it. With the strange horizon, small hills on the "surface" will give a vantage point better than any mountain top on earth. The land will spread out in a broad, up-curving panorama of tightly planted forests, open meadows, small villages, and densely packed columns of farms. The whole interior will be a natural countryside in many ways better than earth's because of the perfect control of the environmental conditions.

There will be constant control over humidity, air pressure, oxygen content, pollution, and day-night cycles. The quality and quantity of food produced by the farms will be carefully regulated as will the numbers and types of plants, shrubs, bushes, and trees. There will be an absence of microorganisms and insects not directly beneficial to mankind and a closed ecology. Air pollution and water pollution will be nonexistent. Almost all machinery functions will be automated.

The starship will differ from a space colony in important respects because it is intended to move away from its 20-year orbit and someday intersect the path of another planet around a distant star system. The central equatorial sea will need a retaining cliff much like that described in Arthur C. Clarke's novel *Rendezvous with Rama. Rama,* was a closed-colony starship that swung through the solar system, presumably by accident; it was a cylinder 30 miles long with a sea in the middle. Because *Rama* was intended to accelerate, the sea was contained on one side by a gigantic cliff more than a thousand feet high. In the case of *Rama,* the retaining cliff was on the "south" side of the cylindrical sea and the ship/cylinder was designed to accelerate with the "northern" pole "forward."

Agamemnon will accelerate from earth during only part of the voyage. Since the starship must arrive at its destination at a sufficiently slow speed to enter the system and explore, it will have to decelerate toward the target. The main method of slowing the starship down will be to perform a 180-degree, end-over-end maneuver so that the engines are thrusting in an opposite direction. Deceleration will also be done using a drag screen. The retaining cliff will keep the central sea from flooding over the landscape of the habitat when the ship is accelerating and decelerating. At $\frac{1}{50}$ g, the central sea would spill over the retaining wall unless it were over 1000 feet high.

From inside, the retaining cliff around the central sea will dominate the landscape. Seen from the top of the equatorial scarp, the water will look like a great *horizontal* curving bay. It will be an illusion; the water will encircle the ship, as will the two retaining cliffs. On a world basically egg-shaped, the perceptions will be visually distorted, and little of it will look "flat" from a distance. Only from a very close perspective will the appearance of flatness be real.

The starship, unlike a space colony or habitat, will not be able to use solar power for energy except when it is in the solar system during construction and on arrival in the vicinity of another star. Scientists who have investigated the idea of space colonies have estimated that more than 100 kw of power could be processed from solar energy for each colony member. People in the highly industrial areas of earth use about 10 kw per day per person, so there would be plenty of power for the starship from solar energy. Unfortunately, for most of the starship's voyage, no solar energy will be available.

If the fusion system on the starship has an efficiency similar to that envisioned by the British Interplanetary Society for its *Daedalus* starprobe, and if about 10 percent of the fusion output were used for internal power, *Agamemnon* would need about 40 tons of "environmental" fuel for a 500-year voyage. Part of the energy would be used to provide artificial light for agriculture, as well as for psychological reasons. If the propulsion system has a better burn fraction in its fusion reactor than the *Dae-*

dalus people have proposed, then the fuel diverted to internal use would be less. The internal-power requirements, in any case, are attainable through fusion designs.

The ship will also require shielding which differs from that of a habitat in orbit in a solar system. Space-colony proposals have involved some fairly complex and as yet not developed technology. On the earth's surface, mankind is exposed to three kinds of radiation: from the soil, rocks, bricks, and other environmental sources; from the tiny amounts of substances which are radioactive inside our bodies; and from cosmic rays which penetrate the atmospheric envelope.

The measure of total dosage of radiation over a given period of time is the "rad." Human radiation produced in the body varies between .04 and .25 rad per year. The radioactivity from the environment varies from a low of about .04 to about 0.9 in some high-radioactivity regions, such as some of the sandy areas in India. It is generally accepted that 5 rads per year will not damage human beings in any way we are aware of now. The dose from cosmic rays varies from about 0.4 rad at sea level near the earth's equator to about .14 rad per year at 10,000 feet in the midlatitudes.* The average inhabitant of earth receives an average annual dose of scarcely measurable size. It is difficult to detect any effect in humans from an average dose of less than 25 rads per year.

Out in space, however, the primary galactic radiation—highly penetrating cosmic rays—is at least 10 rads per year. If this were the only problem with cosmic rays, the shielding for a starship would be minimal. The cosmic rays of space, however, have a property never apparent on earth. Some of the cosmic rays are the heavy primaries—nuclei of iron, carbon, helium, and others. When they penetrate objects in space, they produce

*The earth's magnetic field provides a shield against cosmic rays and the magnetic field's intensity varies. At the North and South Poles, the annual dose of cosmic rays is much higher.

ionized atoms. This ability to ionize is extremely dangerous to human cells. It is also the reason why they do little damage on the earth: our atmosphere is quickly ionized by the cosmic rays, and their energy is lost at very high altitudes.

The heavy primaries were what caused early astronauts to see light flashes, indicating the cosmic rays were impacting directly on their eyes. The astronauts were, of course, outside the protection of both the atmosphere and the magnetic fields of earth. No one knows what the long-term effects of exposure to heavy primary cosmic-ray penetration on a human being might be. It is certain that whereas a space colony might be protected adequately over 20 or 30 or 50 years, *Agamemnon* would have to be protected from all potential effects for centuries, since the crew would be trapped in the starship. Not only could increased cancer rate occur, but so could mutations.*

A space colony could be protected from cosmic rays if the colony were electrically charged to 10 billion volts. Unfortunately, charging a colony with 10 billion volts would also provide an unavoidable attraction for every electron within several hundred miles.† Since an electric shield would be no better than no shield, some complex designs involving the use of electric current to produce magnetic fields have been proposed.

A far simpler method was suggested a few years ago. A space-colony hull could be covered to a depth of 6 feet with lunar soil shipped from the moon by mass-driver. Such a covering would protect the colony from cosmic rays, and the interior-radiation level would then be not much larger than the background radiation in high-alti-

*Genetic damage is not repaired during an individual's lifetime; it is the total dose received, rather than the dose rate which is important.

†Electrons would strike the colony, producing strong X rays, and the electric charge would be neutralized. This assumes that a colony surface—or the surface of a starship—is metallic in nature.

tude cities such as Denver, Colorado. Though the idea is attractive, it is also quite difficult to accomplish, requiring the transport of over 10 million tons of material or more.

If this type of shielding is the most practical at the time the starship is built (it is certainly the cheapest, in some ways), then the ship could be double-hulled. The inner hull would contain the main crew/colony. Around the inner hull would be a layer of lunar soil blown in the way insulation is blown into an attic. The outer shell would be an aluminum sheath surrounding the inner hull and shielding.

With a starship, there is an additional problem in the space environment because the effects of radiation are magnified by velocity. A particle weighing less than 100 mg has enough energy at $0.1c$ to cause a very significant radiation level, a dose sufficient to harm the crew of an unshielded starship. Each molecule of interstellar gas is a hazard with a reasonably high energy level. By far the worst danger associated with the ship's forward velocity would arise from encounters with stray meteoric particles. While the interstellar dust and grains would produce significant erosion on the starship hull and any exterior structures, a meteorite .14 inches in diameter traveling at $0.1c$ has the energy of about 10,000 pounds of nitroglycerin exploded in air.

Agamemnon can be protected from radiation effects and impacts associated with velocity by a shield placed ahead of the starship. While meteoroids probably occur only within solar systems, stray particles, molecules, and interstellar dust and grains will be in the ship's path. Whether there is meteroid-sized matter in interstellar space is one of the questions an ultraplanetary probe would be expected to discover. Erosion on *Agamemnon* from random impact by stray atoms will not offer the only danger. X rays will be produced by impact with hydrogen atoms or molecules, and gamma rays and particle showers will be produced by impact with stray heavy atoms such as carbon or interstellar oxygen.

A heavy shield could be placed at either end of the starship as a sort of "bumper" or a self-propelled shield might be used to travel ahead. At 10 percent of the speed of light a shield at a distance 1000 times its own diameter would reasonably protect the ship. *Agamemnon* would need to be equipped with several shields, and standby equipment if one were destroyed through impact. Instead of being connected to the starship, the shields could have their own propulsion system and be completely computer-controlled from on board the ship. While an impact bumper or a so-called precursor self-propelled shield could effectively be used during a starship voyage, there is no substitute for extensive shielding of the starship itself. But other forms of shielding could cut down on that initially required for the hull.

The starship will need an observatory and a complex one at that. The telescope aboard should be designed with a resolution of 0.1 arcsec, which means that Jupiter-type planets could be seen at 5.07 light years and earth-type planets at 2.5 light-years.* A 0.1-arcsec resolution is about ten times better than that obtained in practice at Mount Palomar, although the telescope would not need to be as large as the Palomar one, because it would be used in space without the problems of atmosphere.

Planets at this distance would appear only as unresolved points of light separate from the central star. Large bodies would show a planetary disk at 40 times the distance to Pluto, about 0.02 ly. Earth-type planets would not show a disk until 20 billion miles. Radio planetary-detection equipment would complement the information coming from the optical instruments. Studies of Jupiter by spacecraft suggest that a starship could detect

*Having a telescope system like this aboard the starship implies that the magnitude limit of the telescope would be 24 and that some sort of occulting disk would be used. It also assumes perfect operation, which is unlikely.

planets by their magnetospheric radio emissions as well as by other types of radio sources emanating from them.

An observatory at one pole will use both optical and radio instruments for navigation. Connected with it will be communications and data-reduction rooms for the unmanned spacecraft that will be sent to the planets of the destination star. The craft will be released as *Agamemnon* approaches at slow speed. The information from the probes will be used to determine the characteristics of the planets and to evaluate them for habitability, as well as to make scientific comparisons with those of the earth's system.

The interior of the starship will be fairly well populated during the second decade. The entire ecological system will be tested by an interim crew. At the same time, the starship will be extremely expensive. The original expedition to the asteroid belt to capture asteroids and the unmanned probes needed to locate them will be costly, as will be moving of bodies to other orbits. Lunar materials, even for a well-developed space society, will not be cheap. To pay for all this, the hypothetical Starship Corporation might gather in one or more new asteroids to mine.

If metal-rich asteroids are chosen, the mining profits can go toward supporting the project during the second decade. It is not inevitable that the balance sheets for the project be completely one-sided. If it stays in orbit long enough as a mining enterprise, with occasional forays into the asteroid belt for a new body, it might eventually pay for itself completely.

During the second decade the interior will be completed. The appointments of the starship may also distinguish it from a space colony as greatly as the retaining rim around the equatorial sea. *Agamemnon* is designed for generations of habitation without failure and without possibility of outside contact. Because of this the residential provisions may be different from those proposed for a space colony.

Aboard the *Agamemnon*, there will be a new society, new concepts of architecture. Much of it will evolve over the years of the voyage. It is difficult to predict what form a starship society might assume; it will probably involve freedom and individuality of expression.

Authors of science fiction and scientific visionaries have speculated on urban planning and high-density housing for closed and open ecologies, near-earth space habitats, and colonies. Some of their visions have the flavor of Orwell's *1984;* some seem like a combination of the Land of Oz with a Utopia conceived by *Trout-Fishing Journal.* Architects usually have a good time with drawing-board concepts, and long before space colonies and starships captured their imaginations, they dreamed up unusual habitats. Alexander's architect, Dinocrates, for example, devised a bizarre plan for Alexander City. Mount Athos nearby was to be carved into the image of a man (presumably representing Alexander). One extended arm of the gigantic figure held a water resevoir to supply the city. The residential areas lay in the figure's lap and on the plain formed by his carved cloak. Roads up the mountain-figure to the city passed under its huge leg. It was a dream never realized, of course.

A city is a unique organism, whether aboard a starship, in a space colony, on the surface of a planet, underground, or beneath the sea. It is, as Lewis Mumford said, "a fact in nature, like a cave, a run of mackerel or an ant heap. But it is also a conscious work of art, and it holds within its communal framework many simpler and more personal forms of art." There have been envisioned cities that stretch completely around a world, offering total urbanization. The world-city has been investigated by some writers of science fiction: one appears in Isaac Asimov's *Foundation*—the world of Trantor, headquarters of the galaxy.

The city and its environment is a frame for human endeavors, a stage for man's thoughts; a stage where both the amusing and the frightening acts of daily lives will be played. And the city is unique to human beings. Insects have organized in semblances of communities, but they are not the same, unless we understand insects much less than we think. The city is like nothing else in nature: it is built solely for the benefit of mankind; it is totally self-serving in purpose. So is a starship.

Two kinds of planners are concerned with cities of the future. One is the group that dreams of a Utopia: philosophers, would-be philosopher-kings, and social reformers, those who do not like what exists now and probably would not like the future any better. In a second category are those whose plans for mankind would permit society to live and develop without being *designed* in any way; genetics, cities, social order, would not be subject to any overall control.

The gigantic interior of a starship, empty except for farms and a central ocean, will be a tantalizing opportunity for both groups. There are almost limitless possibilities. Since all designs for the future express a deep-seated dissatisfaction with the present, the residences and structures will carry the hope of humanity for something better.

Experience in the space program to date does not give a really good clue to the requirements of a starship. Plans for space vehicles carrying up to 24 men for as long as a year provide 300 to 700 cubic feet per man. Those estimates are based more on rocket-launch capabilities than on human tolerance limits for particular periods. Some studies of both actual spaceflight and simulated spaceflight suggest that the confinement of an individual or group in a space less than 150 cubic feet per person results in severe impairment of functioning within a short space of time.

Human space requirements are not absolute. What is normal for individual space has varied throughout history and has generally increased historically. Ample room to live is also a function of industrialization and social affluence. In group situations, restrictions on individual space cause a loss of privacy and an unwelcome enforced intimacy. Although there is little hard data on long-duration spaceflight, it can be assumed that any crowding or confinement to close quarters will produce stress, which will increase as the spaceflight continues. Studies at scientific stations in Antarctica involving small crews of up to 30 men have shown that irritability, depression, insomnia, boredom, social withdrawal, dissatisfaction, and deterioration in group organization and cohesion vary in direct proportion to the degree and length of isolation and confinement.

Isolation aboard a starship, at least isolation from earth,

obviously will be a problem. Even full communication with the home planet cannot prevent the feeling of detachment from the rest of the human race, even though the starship crew would be the size of a small city. Fortunately, the amount of space aboard, per person, can be quite large. The amount of extra materials, construction costs, and other factors involved in building the starship large enough to give the crew adequate room does not present insurmountable problems over a smaller size. There is no point in cramming humanity into space that is minimal according to our present views of civilization and technology.

Cities in space have ranged from Utopian designs to fairly rough dormitories for construction crews depending on the designer's notion of what life should be like in the colony or starship. One wheel design, a moving starship, was divided into small individual cabins where the crew slept from birth to death or until they were awakened at their destination. Paolo Soleri designed a future city around spheres and hemispheres that would hold 15,000 people. The city occupied 7 acres. O'Neill suggested small "villages" for one of his space colonies, a decentralized vision of humanity. He had a preference for hamlets, parks, and forests with lakes at valley ends and at the base of hills. O'Neill projects a pastoral setting: "a house in a small village where life could be relaxed and children could be raised with room to play; and just five or ten miles away, a small city." In his Island One Colony there would be more personal living space per person than probably would be possible in a high-density city of the future.

Some rather bizarre ideas have been advanced for human living—in space or on the surface of a planet: the City of Order, where the inhabitants were designed (via genetic tinkering) to fit the city, and not the reverse; and New York of Brains, a gigantic cube filled with millions of 10-inch individual cubes, each of which held a human brain; and Conical Terraced Cities, with circular terraces one above the other, with the population receiving orders and instructions from several forms of "Mr. Big" in the penthouse. NO such fantasies will be realized on *Agamemnon.* There will be residences, shops and offices, schools and hospitals, assembly halls, open spaces, perhaps some areas designed for light manufacturing, and space for storage, mechanical subsystems, and transportation. Some of the clues to the amount of room needed for these activities can be obtained from what exists, at present, on earth.

Aboard the starship the space requirements might be substantially different, but on earth 1 shop per 100 population seems to be about normal. If it is assumed that a total of 37½ percent of the starship's children would be attending school, space of about 10 square yards per student would be necessary. Since the *Agamemnon's* crew are more likely to be childless at the start, an overall planning figure of 10 percent of the total starship population can be used in figuring areas for schools.

Based upon the average number of patient days per year for an earth population similar in characteristics to those aboard *Agamemnon,* a hospital of at least 26 beds is required. But because space travel may expose the population to increased dangers, a 50-bed hospital would be more appropriate. The hospital would include areas for administration, surgery, obstetrics, treatment, diagnosis, nursery, and emergency.

The average amount of space in the United States devoted to parks and other open areas is about 18 square yards per person. A starship, however, has agricultural areas that can serve double duty in planning. For 10,000 people there should be, on the average, about 1.5 square yards per person for churches, meeting halls, and theaters. For recreational areas, about 1 square yard per person would do.

A total of 500 square yards for each of 3000 families

would be needed for mechanical subsystems such as communication and distribution. The same 3000 families would need 40,000 square yards for waste and water treatment and recycling and about 1000 square yards for electrical supply. Since the starship is fairly densely populated, the major source of mass transportation would be a moving sidewalk or small rail system.

The starship has some possibilities for changing the apparent population density that are difficult to use on earth. There can be stacked living areas in widely spaced levels; staggered working and sleeping schedules can be designed for optimum use of recreation areas; and there can be much doubling-up in use of facilities. The residential spaces aboard could be scaled from a Paolo Soleri design called *Arcube. Arcube* was an exercise in extreme, high-density construction that depending upon your viewpoint, looked like two A-frame dwellings base to base or a cube resting on one corner. The original model was to be 5000 feet high, covering only 346 acres. It would hold almost half a million people. There could be three or four small *Arcubes* housing the majority of the population. Situated around the cube-cities would be meadows, forests, a circle of farms, and the circular ring of the equatorial sea.

Imagine floating cities for the population. The cities could hover above the curved inner surface of *Agamemnon* and change locations at will, providing some variety, which would be appreciated during what might become a very long lifetime. Another scheme would be to build conveyor-belt cities which would differ because of their location. One at the "north temperate zone" might have near-zero gravity, one south of the central sea might have half earth gravity, and one at the opposite temperate zone might have Earth-normal gravity, for the "conservatives."

Since there are tasks which can be accomplished in near-zero gravity which cannot be done in high gravity, the cities might develop vastly different styles in living, despite being relatively close together in physical space. In this sense, they might more closely resemble city states. Since there is much about zero gravity that is sensual, from the viewpoint of human experience, perhaps the near-zero-g city would be a hedonistic Utopia, Bacchanalia gone mad. The middle city would be inhabited by moderates in all things. The remaining city might be a depressing combination of religious fervor for Earth and police-state ideology.

When Sir Thomas More used the word "Utopia" in 1516, it could have been a mocking name for a place that was nowhere or it could have meant a "good" place. The interpretation lies in how you read your Greek. Utopia is now such an emotion-laden term that it cannot be used in a neutral sense. All of social science and reform is not Utopian, nor is any culture necessarily "free" in a realistic sense. *Agamemnon* will probably be a community of joys and sorrows not fundamentally different from those we know today. People adapt to unusual or novel circumstances, but they do so in thoroughly human ways. And when communities are severely disrupted, they die.

The environment of the starship will dictate, to some extent, its interior arrangement and design. The mild weather will eliminate any need for roofs or walls except for privacy or for aesthetic reasons; nor will the architecture necessarily isolate the inside of a dwelling from the outside. Permanent buildings can be optional; they might be modular and changed to suit individual taste or the expressed taste of a community; they could be changed for variety every decade.

It is no more desirable to make the cities, dwellings, and apartments uniform than to make them flexible and easily changed. The idea that sameness is an expression of unity is unscientific, unbiological, and contrary to nature. The most important element of the interior design of the starship must be freedom—pure, unadulterated, unplanned, and generally unsupervised freedom. Flexibility offers freedom.

Since the starship crew will be close to their natural environment, the interior design might reflect a direct visual, olfactory, and even tactile relationship with "nature." This atmosphere could be attained by movable walls, large windows, low and high partitions, natural surfaces as flooring, gardens both interior and exterior, and other variations of living in a natural surrounding. Trees growing outside windows give the impression of a surrounding forest, which can be heightened by filling part of the interior with miniature landscaping. Planned landscapes can be changed to simulate the effect of changing seasons as they occur on earth. If this idea were adopted, the process of changing the interior to correspond with seasons would involve a large part of the population; everyone would plan for it, take part in it, and the simulation would make the change "real" for the inhabitants.

It is almost impossible at this point to speak in *concrete* terms about the starship environment. Perhaps it is best left up to the crew who have to live with it since all our theories about human behavior in confined environments have been earth-based, with the exception of the data from *Skylab, Salyut,* and simulated space missions of relatively short duration. The theories we now hold are so couched in the attitudes of an earth-based society and surrounded with a multitude of cultural, political, and historical traditions and customs that objectivity is difficult.

Agamemnon, after a generation, might look more like the Hanging Gardens of Babylon than a starship. It might be as sterile-looking as a psychiatrist's office or it might resemble nothing we can readily imagine today.

The starship will not be an "all play and no work" society.* If it were, it would probably fail. But leisure time will be a benefit for the crew, and entertainment

*Il n'y a pas de sot métier! (There is no such thing as fool's work.)

will be important aboard the starship. In addition to some of the traditional earth-style forms, there will be sections of no weight near the poles and along the central axis. Only slightly distant from these points, the gravity is small. A low-gravity swimming pool would be nice, as well as a low-gravity dance floor.

A swimming pool at $1/20$ or even $1/50$ gravity will remain within the confines of its shape. In the case of a pool aboard *Agamemnon,* it would probably be cylindrical, like the central sea. A cylindrical swimming pool would change its aspect depending on the viewer's perspective: the pool side from nearby would appear as pools usually do on earth, but as you look up, the pool would curve and eventually be overhead. From poolside, there would be people splashing and swimming around to the right, left, and above. While pools aboard space colonies will probably have been common long before a starship is built, the unusual effect would still be astonishing.

Waves in a low-gravity swimming pool would also be vastly different from those on earth. The waves would sweep much higher, and substantially slower. A handful of water would retain a reasonably cohesive shape from surface tension in low gravity, and a water fight in a low-gravity swimming pool would more resemble a fight with water balloons on earth.

The people aboard the ship could swim like dolphins in the swimming pool. Though the act of swimming would not be different from that on earth, a crew member could speed along underwater and then rise to the surface and execute a graceful flight over the water in the low gravity. If there were a real dolphin aboard, it could perform antics the Disney people would never have anticipated. It would also be possible, under certain conditions, to perform the miracle of walking on water—but aboard *Agamemnon* it would not be a miracle but rather commonplace. Diving in low gravity would also produce some amazing new sights.

The human body can be moved with exquisite grace and motion in various art forms on earth, and in low gravity the possibilities are increased dramatically. Flesh moves in slow motion; all movement is in slow motion. A leap could last for minutes, a spin for as long as a dancer could stand the centrifugal forces. Choreography would take on a new dimension. A dance floor or theater up in the low-g areas would provide a delightful leisure-time activity. It is also possible, in the low-gravity areas, to "fly" in air on spindly aerodynamic forms: kites, human airplanes like hang gliders, and small powered airplanes somewhat resembling the Gossamer Albatross. The recreation areas of the starship would be like nothing ever seen on earth, with the strange swimming pools; great, slowly swirling ballets; and people dipping and gliding like dragonflies about the landscape.

Like other societies in human history, both minor and major, *Agamemnon* will have its share of hedonism. Arthur C. Clarke once remarked that weightlessness would produce new forms of erotica, and in this prediction he was probably correct. Sex and the zero g should relegate the *Kama Sutra* to the level of the *McGuffey Reader.* Somewhere up in the low-gravity sections of the starship there should be a series of exquisitely and specially equipped rooms.

Perhaps one of the most distinguishing features of a starship society would be the arts, particularly the art of story-telling: the second-oldest profession, the court poet. The minstrel may replace the block leader; the dancer may replace the ward heeler; the poet may replace the politician. A long stretch of human history has gone by since the time when a king dared not insult a poet. The last king penalized for the deed was Breas of the De Danann, a people of Ireland who were eventually defeated by the Firbolgs.

Breas's greatest shortcoming, according to legend, was in the realm of the arts. During his reign there were no festivals for bards, no satirists or jugglers, no pipers or harpists. The arts were banned from the royal presence. His reign might have survived a lack of prosperity, but not a disdain of the arts. He made the mistake, unpardonable in ancient Ireland, of insulting a poet. The great poet Cairbre, while visiting the king, was relegated to a garret room in the royal house and given a few cakes of dry bread and some ill-bred wine for company. Infuriated, Cairbre composed a withering satire on the king; the people, upon hearing it, revolted; and Breas became an ex-king of Ireland.

MONGRELS ARE THE SMARTEST DOGS

"When ships to sail the void between the stars have been invented, there will also be men who come forward to sail those ships."

JOHANNES KEPLER, 1610

Spacecraft flight crews, from *Mercury* through *Skylab*, have been small, quasimilitary, and homogeneous units. The crews have been trained in the same way over the same length of time with the same tools. There was, as the press was fond of pointing out, little difference between the astronauts; nor was there much distinction to them. That some were better-suited to particular tasks than others was evident, but overall they were faultlessly and annoyingly alike. There were all test pilots. All had military backgrounds. All were men.

The astronauts could survive centrifuge training to 20 times earth gravity—20 g—until the victim's capillaries were bursting, and teeth were the only recognizable facial feature. They survived being shut up in ovens at 140 °F for two hours at a time; they were thrown into ice packs to test their reactions to cold. They lived in burning deserts and hostile jungles as training. Most of all, they could relax inside an aluminum can moving through a vacuum intolerably dangerous to them as human beings. They adapted to zero gravity.

In truth, they were supermen, trained for eventualities beyond belief; exploring unknown territory; able to cope with any stress, any pressure. They were watched, punched, and probed like laboratory rats and with, sometimes, as little courtesy as that given to rats.

Machines have eliminated the need for an astronaut as he was originally conceived. The space shuttle is often referred to as a "shirtsleeve" design, meaning space suits are not required inside. There are jobs in space that can be performed by those not trained as military pilots. With the space shuttle, there are, for the first time, distinctions that are not arbitrary between crew members. There are command pilots, copilots, mission specialists, payload specialists, and someday there will be "passengers." The pilots are still highly skilled, superbly trained. The others are not. Women have been included at last.

As we move into space with space stations, construction bases, large space structures, lunar bases, and space colonies, the crew requirements will become increasingly varied, and so will the problems of dealing with people.

There are some who think living in space will, in some magic way, produce a better human being, free of those characteristics we do not like about the race. They see it as an opportunity to sweep out the old and pour in the new. Terms like "cultural engineering," "functional integrity," "common earth-type deviance," float with uncomfortable frequency from papers devoted to decrying the lack of "social planning" for space.

Gerard O'Neill, who sparked much of this dialogue in what is loosely termed "space sociology," was extremely careful to stay away from making judgements about the social structures of his colonies. But it is possible to read into O'Neill's proposals certain values: there is no mention of conflict, hate, depression, or stress on personal, economic, bureaucratic, political, or ideological levels. His colony is a wonderful place to live. It is a place designed for a "comfortable life ... enlarged human options ... breaking through repression."

The people who worked on the NASA Space Colony Study in 1975 gave some consideration to possible types of social organization. They envisioned one community that had majority rule as the basis of democracy, with competition as the basis of "progress." The community was described as hierarchical and homogeneous; differences were considered accidental or merely inconvenient, and diversity would be considered abnormal and undesirable. Such a society might feel that differences create conflicts and would prefer to organize everything and everyone for maximum efficiency.

Another community foreseen in the 1975 study was one in which independence is raised to a high level of virtue both for the individual and from the viewpoint of the group. Such a society would consider self-sufficiency as the highest form of human behavior. A community like this might be loosely based on the precepts of individuality and creativity discussed in the novels and essays of Ayn Rand.

A third social order considered was one in which heterogeneity is considered a source of flexibility, evolution, and survival. Rather than thinking in terms of the greatest efficiency, a society like this might think more in terms of choosing and matching people for a reasonable result. A society so structured might consider competition useless but cooperation very helpful. The overall idea is more Eastern than Western, somewhat like a Japanese flower garden: a harmony in diversity.

Recent experience in space and in simulated space environments shows what we might expect from mankind in the new frontier. The third crew of *Skylab* began, after a few days, to respond to communications from Houston with a blend of hostility, ridicule, and exasperation. They were different from the first two crews in two ways: none of the third crew had been in the navy, where shipboard experience helped, and some were from a science background instead of military. Their irritation aboard *Skylab* ranged from complaints at the level of "There's no light in the men's room" to near-serious dissidence. The third crew seemed to be occupying a space station substantially different from that inhabited by the first two crews. They did not like anything about it most of the time.

One of the reasons for the discontent of crew three, aboard *Skylab,* relates to some of the previous discussion about planning and how humanity in small groups and large masses reacts against it. NASA had bugged *Skylab* as thoroughly as they would a suspected den of spies. The astronauts were unaware of this and, being human, said and did some things that violated some of the thousands of rules established by Mission Control, and NASA planners. The astronauts were caught by the ground administrators and publicly reprimanded, something never before done to an astronaut.

The result of the conflict was classic human "deviant" behavior; the astronauts felt betrayed. Conversations with the ground were tense and argumentative. The dissatisfaction was aggravated by the lack of communication during more than 25 percent of the orbit, when *Skylab* was "blind" to ground stations because of its high orbital inclination. The number of small and inconsequential errors made by the crew slowly grew. They dropped behind schedule. Mealtime, technically a psychological high point of the day, became increasingly Rabelaisian. The men refused to exercise after meals. "We're just being driven up a wall," one *Skylab* crew member said during the third mission. In all of the history of the American space program, no astronauts had ever talked to Mission Control the way the third crew of *Skylab* did.

The affair of *SMD-III* is even more foreboding. The spacelab, designed by the European Space Agency, is to be launched as a payload of the space shuttle and is not scheduled to fly in space until the mid-1980s. Its program was tested in simulated flight at Johnson Space Center from May 17 to May 23, 1977, in what were called the Spacelab Mission Development Tests. The simulated flight was complete with a seven-day trial of payload experiments.

The crew of *SMD-III* consisted of an astronaut mission specialist and two payload specialists. The flight crew who would have piloted the shuttle into orbit were not included in the test. The reports show that conflict arose between the payload specialists and the astronaut specialist during training and during the simulated flight. One of the most common conflicts concerned authority over the payload experiments. Who was responsible: the astronaut specialist, the payload specialists, or the principal investors (called PIs), who were not in "flight" and were not employees of NASA? Had it been a real mission, the flight crew would have been vying for power, too. A full spacelab mission would involve a commander/pilot, an astronaut mission specialist, two or more payload specialists, PIs on the ground, and even a "passenger."

Partly as a response to the inherent potential for conflicts aboard space shuttle, NASA issued a memorandum in 1980 that gave the spacecraft commander the same rights, privileges, responsibilities, and legal power as the captain of an aircraft in flight.

The elitism of the Astronaut Corps, partly self-generated and partly unavoidable because of status and training, was what prompted so many scientists to refer to them as "bus drivers." This was caused, sometimes, by the inability of the scientists, who had university backgrounds, to become part of a team effort in the quasimilitary style of NASA. Some of this very early "conflict" in space shows that human beings have not been changed by the new frontier. Alienation, pressure, the usual emotions—love, hate, envy, pride—and the range of human reactions are not basically altered by the space environment.

Agamemnon, during the period in which it will be essentially a colony of earth, will not differ socially from other colonies. While still in orbit around the sun, it will retain the social and physical intercourse available to other colonies, and its authority will be earth-derived. If not, then authority will be derived from a space station or the space industry. When the ship journeys between the stars, relationships will be different. What planners will have to do is try to anticipate, with little available data for support, what can go wrong with the social order of the starship.

In earth or solar orbit a social breakdown would be drastic, but hardly fatal. In space the social order must be stable, or the trip will end. Stability does not necessarily imply rigidity; rigid societies are often more unstable than more flexible ones (the same is true for individuals).

Preplanning the social order of the starship crew may seem an attractive enterprise, but the chance of success is poor. No amount of social planning can predict conditions aboard *Agamemnon* beyond a few decades into the voyage. Murphy's Law may be operable: whatever can go wrong, will. But there are some areas of planning that will help make the trip easier for the crew, such as designing a physical environment as desirable as possible.

On earth, isolation is never complete. Though a sensory-deprivation tank comes close to duplicating psychological isolation, the individual still knows that just outside the tank there are people, and outside the building is sunshine, or rain, or trees and grass. Aboard a starship in space, the isolation would be total. Just outside the hull, only a few hundred feet, will be empty space; people will be billions of miles and dozens of years away from anything familiar. Outside pressures, once the ship is beyond communications distance from earth, would be nonexistent. All pressures would come from inside. In this respect the isolation might be beneficial; there would be no one to criticize the changes that might be made aboard. The adults would never come home from the party to confront the children with the mess they have made of the house.

Newness would be the key aboard *Agamemnon.* Variety and flexibility in home design and cities should be extended to jobs and hobbies. Most of the crew would hold more than one job, which would give some relief from routine. Preferably the jobs would offer different challenges or, at worst, be only slightly similar. Skilled and unskilled labor could be alternated.

Whether the dream of the starship voyage will be sustained throughout the trip or become only a vague collective memory reinforced by a massive selection of videotapes or their equivalent in a learning center is an unanswerable question. For succeeding generations aboard, the survival of the community and its continuation will be the prime motivation. The voyage to the stars will be secondary. In certain respects, viewed from the day-to-day, year-to-year viewpoint of a crew member, the voyage *is* of secondary importance. Amid the daily uncertainties of life, the important considerations will be friendships, love, happiness in the job, advancement, birth, death, and all the usual human concerns that occupy individual minds. Keeping the inspiration of traveling to another solar system alive may be difficult.

When the great projects of antiquity, most of which lasted for generations, were in progress, the first generation might conceivably have been sustained by the process, but succeeding generations were probably not. Though it is romantic to conclude that many dreams have been kept alive by the devotion of large masses of people, such is not necessarily the truth. They are often sustained by a small minority who have held or could influence political power. Whereas dreams move some people, law often moves the majority.

The pyramids were built by the Egyptian populace because the ruling government commanded and the people obeyed. It would be difficult to find an individual Egyptian crawling up the steep incline of a pyramid under a fiery sun who would be lucid and thoughtful about the architecture of pyramids or the future heavenly well-being of a dead king. And though there is reason to think of the pyramids as public-works projects that absorbed surplus labor, the ancient Egyptians would probably not be disposed to discourse about this theory of government economics.

Much the same could be said of Stonehenge, the statues on Easter Island, and most of the wonders of the ancient world. They were conceived by dreamers of a kind, and the vast majority of construction was done by laborers under threat—overt or implied. The Egyptians, however, did have a strong element of religion (Pharaoh-worship) in the communal labor of pyramid-building, and there were religious elements in building the monuments of Polynesia and Megalithic Britain. Perhaps our romanticizing of the wonders of antiquity is much like our romantic view of chivalry: the imagination of a later century imposed on an earlier one. Chivalry was an invention of the eighteenth-century Romantics pasted over the dreary reality of feudal twelfth-century history.

The wonders of the modern world are only slightly different. It is no longer considered proper to threaten people openly into work. The threat is more subtle; it is economic. There was pride in building the Empire State Building, and motivation, but how effective pride would have been without the steep pay scales in the middle of a depression is subject to question. The noble idea of an Empire State Building might not have sustained the workers for a decade. John Jacob Raskob might have kept the idea alive while he lived; and perhaps the architect, William Lamb, might have kept working for his dream. Humanity has incredible inertia, which is moved for good or evil by the individuals who can inspire others to work to realize their dreams. Nevertheless, the crew of the starship *Agamemnon* may be somewhat different in the way they approach the gigantic project of traveling to the stars.

A large portion of the crew will, no doubt, come from the developed areas of space: space stations, lunar bases, space colonies, etc. They will have already experienced the environment and isolation of space; and they will all be volunteers for a dream they all would presumably share. The tolerance of isolation, stress, and other undesirable aspects of any space mission is much greater when there is a shared and important goal.

This has been demonstrated in a number of studies of working environments in which, as it turned out, what often mattered to people was that they felt a part of a special project or were in some way unique. The idea that the starship crew is another form of "chosen people" has obvious applications, and it may be that single psychological motivating factor that will provide the cohesiveness necessary in the crew to survive generations in space.

If a large majority of the crew are selected from the

space-based communities of the future, who will they be? Will they be technicians? Military people? The adventurous and the unsettled? Who of us would be willing to travel imprisoned in a metal-and-rock cocoon for decade after decade with only an artificial sun for rhythm, no moon at night, and the breath of a neighbor in next week's breakfast? No matter what magic is done with design and imagination, *Agamemnon* will be confining.

The members of the crew will come from everywhere. There is much more room aboard than the average human gets in Hong Kong or San Francisco or Manhattan. Humanity has existed for hundreds of years in cities where the available space for each person was far less.

The people who will climb aboard will be like you and me—and our friends, your wife, her sister, the man next door. There will be those to whom adventure is life. There will be those who have nothing else to lose. Some will be motivated by the spirit, some by the intellect, some by a desire to shape a new direction for humanity. Except for a few who dream—the technicians, the scientists, the captains in their robes of faith; those who will make the decisions at the right time, push the right human buttons at the appropriate moment—the rest of us will run the starship in much the same haphazard and convoluted ways we always have conducted our lives.

Among the breeds of dogs are the beautiful ones, and the talented, valuable, pedigreed, long of limb or bobbed of tail. Others are nervous, moody, savagely aggressive, docile. Some you can hold in the palm of your hand. Some are heavy enough to sink a small boat. Each is in its own way unique and specialized—and vulnerable.

They are the elite of the dog world: fawned over, photographed, pampered. They have to be protected. Mongrels are the smartest dogs. They know how to survive.

12 LAST TANGO AT THE SUN

"If you don't rehearse over and over, you're going to be surprised in space. And the surprised man, out there, is the dead one."

RAY BRADBURY

Once *Agamemnon* is launched, the margin for error throughout the entire ship is frighteningly small. One of the keys to ensuring its survival is the selection of the crew—the human element. The other is the quality of its vast array of complex systems and subsystems: the mechanical, electrical, and nuclear "crew."

The old science-fiction idea of The Computer makes about as much sense as a nineteenth-century factory in which all the motive power came from a single steam engine and was distributed by an army of belts. Modern and future computer technology, like modern factories, will use distributive power: many small motors or computers with specialized tasks giving power or information and decisions as needed.

The computer survivability rate aboard the Mars *Mariner* unmanned spacecraft of 1969 was 22.5 percent over a flight time of 10 years. While later spacecraft were an improvement on this, rates were variously estimated between 35 and 70 percent, even with triple and quadruple redundancy. The idea of redundancy is not new to computer technology, but the increase in survivability over time is not greatly improved no matter how many redundant systems are used. In 1977, for example, the U.S. Air Force estimated that a quadruple-redundant system aboard unmanned space vehicles would only have a 44-percent chance of survival unattended and without degradation in its performance. Equipment with this endurance level would not be adequate for a starship.

It is far more likely that a self-repairing computer would be used, coupled with a trained crew of technicians. The concept of a fault-tolerant computer or self-healing machine has been around since von Neumann introduced it in 1948; it is why there is an IF statement in Fortran. Self-repair means that the computer is capable of detecting, isolating, and bypassing its own faults and providing undegraded operation.

Fault tolerance is achieved by a combination of computer architecture, logic design, and software. The machine is designed to employ redundant elements or modules. Its logic is designed so that there is no possible single point of failure that can result in a complete loss—a down computer. A recovery program in the software allows the computer to be reconfigured if there is a component loss, power failure, or excessive radiation.

The whole process is watched over by a special control unit. Modern fault-tolerant computer systems designed for space vehicles have a survivability rate of 95 percent for missions of 5 to 10 years, and they are being improved constantly.

Flexibility of the computer systems aboard *Agamemnon* goes along with self-repair. Adaptive-learning computers such as those controlling the enclosed environment and ecology could modify their operating hardware as new data were received. A system with rigid goals might be unable to react to hundreds of unforeseeable situations. Flexibility is, and has been, a part of modern computer technology, and the future is sure to see more improvements.

Over many areas of the starship, specialized computers will do the work, such as the one controlling the navigation. The use of supercomputers for specialized tasks has been proceeding rapidly in the last few years. A highly specialized supercomputer that will be operational in 1986, for example, can assist in developing and testing new aircraft designs and other flight vehicles and can do general research in fluid flow applied to meteorology, gas dynamics, and computational chemistry.

The capacity of computers for storing, processing, and displaying information has been greatly improved through the use of microelectronic devices called microprocessors. They are the equivalent of the central processing unit in a small computer. They are fabricated on a thin slice of silicon called a wafer, and an individual device is called a chip. A chip for an Intel 8085, a general-purpose machine that can execute 777,000 instructions each second, is only .164 inch by .222 inch.

The progress of miniaturization has gone so far that the power of a large main-frame computer can be packed into three tiny semiconductor chips. These three chips represent, among them, 220,000 transistors. The performance is well up to that which might be required aboard a starship. Using Defense Department ADA language, three chips could multiply 32-bit numbers by 32-bit numbers in only 6.25 microseconds. The densest chip ever made had 450,000 transistors and could multiply a 10-digit number by a 10-digit number in 1.8 microseconds.

Most of the technological achievements of the past ten or twenty years have occurred because of microelectronics, of which the development of the microprocessor is a part. Small and reliable sensing and control devices are the basic elements that allowed the exploration of space to begin. A future space society will have the advantage of more possibilities, if not in miniaturization (there seem to be limits) then in increased ability. Of all the giant components of a starship, the computers available now come the closest to what we'll need. We could design the system today with a reasonable chance of success.

The computer system of *Agamemnon* might have two separate functions: one segment would operate in real time for day-to-day decisions; the other would deal with the overriding philosophies, goals of the mission, strategies. This kind of dual computer is called "on-line" and "off-line" processing. Another of the supercomputers of the type used for navigation would work in much the same way the systems at Jet Propulsion Laboratories are used for incoming information from spacecraft.

When the starship nears the target system, information on the planets orbiting the star will be coming in from the observatory instruments. A planetary search computer will evaluate this information and also compare the information with any available from previous starprobes launched from earth. When this preliminary information has been evaluated, the planetary computers will program a series of unmanned probes launched from *Agamemnon.* The computer system will handle targets, flight, and data processing from the probes independent of the machines controlling the starship's navigation, course, and environment.

How far miniaturization of data processing might have progressed by *Agamemnon*'s time we can only speculate. Using single molecules is an unlikely prospect. Even existing microdevices have problems associated with stray radioactive particles at the low-energy quantities that define a "bit" of information. There is also the problem of "shot noise" due to the quantization of the energy.

The basis of the ship's computer systems may be the use of cellular organization. There would be thousands of small, identical modules consisting of micro (or perhaps smaller) processors. A group of cells would act as an array processor. The effect would be an analogue to the human brain, the failure of even a large number of cells would result only in a decreased efficiency in performance. The human brain can withstand a relatively massive degradation of cells before more than a few percent of the brain's operating power is affected. The computer data might be stored on the surface of a spinning disk as a minute magnetic region and be written and read by a laser.

Communications will be handled by another complex subsystem of the starship. The receiving link in the solar system will have been built for reception of information from ultraplanetary probes and starprobes—perhaps doubling in use as a giant radio telescope—located on the moon. Although current receiving links could be used for communication over interstellar distances—the Arecibo dish or the VLA in New Mexico—the relative ease of building large structures in space will eventually dictate the choice of one or more giant antennas.

For starship communications, an array in space can be built of almost any size. The advantages of the array design are cheapness and rapid increases in sensitivity. The array, used in radio astronomy, is a series of small dishes connected to a computer that "assembles " the dishes electronically, as if it were one enormous unit. Until the VLA (for "very large array") was built in New Mexico, the largest radio telescope built specifically as an array was in Cambridge, England. It was 3 miles long. The VLA has 27 antennas arranged in a "Y" configuration. When the whole array is interconnected by computer, the effect is the same as a single dish with a diameter of 17 miles. Several specific plans, such as Project Cyclops, for receiving extraterrestrial communications could double as an earth link for something like *Agamemnon.*

A starship might transmit by laser link in the near infrared. The designers of the BIS have favored a laser with yttrium aluminum garnet crystal lattice structure doped with neodymium for *Daedalus.* This YAG system is basically within current technology, in keeping with the idea of using only current or near-future developments for the BIS starprobe project. By the time *Agamemnon* is constructed, the way may have been paved for lasers that would far surpass anything we can contemplate now. The YAG system would be good for distances only up to 2 light years. Future designs for a starship might involve very high purity (and clear) beryllia doped with iron. A beryllia laser system would be extremely simple and almost industructible.

The reliability of communications hardware is good, and it has been tested for decades. Most of our television, telephone, and radio satellites have had an excellent record of dependability. While a system for a starship is still a future development, it is another one of those parts of *Agamemnon* which is close to being available in our own time. With appropriate redundancy, a link with the starship should last for centuries. Some of the telephone equipment now in use, though not quite the same thing, has been buried under city streets for more than a century, and some of the Transatlantic cable equipment is decades old and still functioning underwater.

The greatest difficulty in the link with earth will be the time of transmission and receipt of message. Radio, television, lasers—any communications link we can imagine is limited by the speed of light. As the distance of the starship grows, so will the delay in sending and receiving. While there was less than a 2-second lag in communication with Apollo on the moon, the lag for *Agamemnon* will start with its construction. From an orbit near the moon, the amount wouldn't be great, but if *Agamemnon* is built out near the asteroids, the transmission-reception gap could be over 25 minutes to earth.

When the starship departs the solar system, the steadily increasing distance will make communications lengthy: at 0.1 light year, a message would take 36 days to reach earth and another 36 days for an immediate reply to reach *Agamemnon.* At a distance of 1 light year, a two-way communication would take 2 years to complete (1 year to earth, 1 year to the ship.) Eventually, communications with earth will be a very minor part of the culture aboard *Agamemnon,* and the link will be infrequently used except for scientific reports.

What the starship really needs is a version of the old science-fiction "sub-etheric radio" which could transmit at high data rates, greater than light speed. Of course, no ideas have come forth on just how one actually could be built. Looking over the possibilities is interesting, though. Perhaps *Agamemnon* would communicate with earth by tachyon beam; the beam would travel at high multiples of light speed, and the link with earth would be kept at the level of daily communication. Perhaps a way will be found to aim some kind of particle beam into *nonspace*, and it would arrive at earth instantly or with a tiny lag. That would be odd: could the particle beam go where a starship, also composed of particles, could not? We have, in some ways, only scratched the surface in our search to discover how the universe works.

Navigation aboard the starship will be dependent on accurate data concerning stellar distances. An invariant coordinate system can be defined by distant quasars. It will be in respect to the quasars that all stellar motions and positions will be measured. Such a system is not in use today but it is likely to be available before the *Agamemnon* is built. Radial velocity and distance of the starship can be provided by the solar-system communication link in much the same way that the same information is used for unmanned spacecraft today. The distances will be much greater, but the principle is the same.

There is the possibility of "very-long-baseline interferometry" (ground-based) as a data type. Interferometry is a method of measuring tiny angles by using the principles of interference. In radio astronomy it employs two or more antennas as widely separated from each other as is practical; each receives radiation from the same source connected to the same receiver. The advantage is that interferometers reject background noise. Their basic disadvantage is not particularly relevant to a starship; they are sensitive only to radio waves from sources of very small angular size. Inertial navigation systems could also be developed.

Any of the possibilities or a combination, could eliminate some of the interstellar-navigation problems. During interstellar flight, stars appear to shift because of their proper motion, aberration, and parallax. The proper-motion shift is the same as it is seen from earth; the star's motion in the galaxy will show up over time and the coordinates will change. Aberration is an apparent motion of objects caused by the observer's velocity.

The forward motion of the starship also causes a Doppler effect in stars viewed from the control room. Stars seen from the direction forward seem hotter, brighter, and closer together than they do from earth. Those to the rear of the starship seem cooler, dimmer, and farther apart. The shift in parallax is an apparent motion of an object toward the original position of the observer—caused by the displacement of the observer from his original position. Corrections for these relativistic effects on stellar appearance and position are readily computed and are straightforward.

Part of the navigation equipment aboard the ship will be, of necessity, on the exposed outer hull. The impacts from the interstellar medium and small particles with high energy will be damaging to them. The radio and optical navigation and science telescopes could be placed in shielded locations, but this adds complexity. They could also be relatively exposed, but used only occasionally so they would not be made useless by interstellar erosion at the end of the voyage. An alternative might be to use large, shielded mirrors on the outer hull, which would reflect light beams into the interior, where the observatory would be located. Even if the external mirrors eroded eventually, they could be replaced by spares carried aboard the ship, or they could be remanufactured. The main working components of the observatory would be unharmed.

With self-repairing and fault-tolerant computer systems, highly sophisticated laser communications equipment, and sensitive optical and radio navigation systems, how reliable could the starship be made? Could it be expected to complete the voyage, which might last 500 years or more? The answer is that it is likely that it can be made very dependable as a *unit.* The reliability problem may not be as formidible as it might seem. Testing and retesting in the orbit of the starship will be one way to deal with the problem. And it is worth mentioning that the best Apollo mission, built perhaps centuries before a starship might be built, had an on-paper reliability of 99.9999 percent. It had but two small defects in millions of components. Of course, there was *Apollo 13.*

But the best reliability insurance will still be the presence of humans. Computers, regardless how "intelligent," are a poor substitute for human drive and leadership. It is one thing to ask a computer to keep the census, allocate pig A to farm B, or to measure the diameter of a rocky planet around another star, and quite a different thing to ask it to exercise judgment, discretion, insight, and hope—that most human of all functions.

When the starship is nearing completion—at least as far as the interior is concerned—there will be two major hurdles remaining: the propulsion system and a classic shakedown cruise around the solar system. Propulsion is vital, because it will determine, along with the chosen destination, the length of the starship's voyage. And as earlier discussion showed, propulsion, when it comes to the distances to the stars, is a thorny problem.

It would be nice to speculate that antimatter would be available to power a starship. Using 4 tons of "reaction mass" for each ton of payload with the reaction mass heated by matter-antimatter destruction would be convenient. But there has been very little real progress in developing a method for propulsion by antimatter since it was first theorized. It is possible to project that solar-array-powered laboratories may be producing quantities of antimatter in the next century, but there is no reason to suppose that it will be enough and it might not happen at all. Antimatter is extremely hard to handle, and though it bears some relationship in this respect to many other substances which mankind has tamed to its use over the centuries, it is also substantially different.

Fusion still seems to be the best prospect for a starship, and a fusion propulsion system will probably propel *Agamemnon* out of the solar system toward a distant star. There are indications that it might be available to the world of space bases, colonies in space, and the starship builders.

In 1976, Gregory L. Matloff of the Division of Applied Sciences of New York University studied the problem of moving a large mass at a velocity which would make starflight voyages possible within reasonable time spans. Matloff assumed the same fusion-propulsion performance as the engines designed for Project Daedalus. He believed that level would represent the best propulsion ability available, even after another century or more of fusion research. The engine, burning deuterium and helium-3, has an exhaust velocity of 0.03*c* and Isp 1,000,000 sec.

Matloff made the assumption that one of O'Neill's early space-colony designs would be used as a "space ark" and that the amount of fuel available for the starship would be the same as that used by the *Daedalus* starprobe for acceleration. He theorized that the ship could use electrostatic means for deceleration, and it could reach 0.002*c* using planetary rebounds—the "gravity assist." When all of his assumptions had been made, he concluded that the space ark would have a cruise velocity of 0.01*c*. This would allow flights of 400 years' duration to Alpha Centauri, 600 years to Barnard's Star, and a trip of 1100 years to Tau Ceti. His basic assumption was that space-ark starflight was possible given the advances implicit in the development of space colonies and the building of something like the *Daedalus* starprobe.

Though Matloff's proposal indicated that such a development was possible, there are numerous improvements that research into fusion and rocket technology over the next century can bring that would increase the chances for the *Agamemnon* voyage. The best fusion engine yet developed on paper has Isp 1,000,000 sec, the limiting figure Matloff used. But the theoretical limits for a laser-fusion design is two and a half times that: Isp 2,640,000 sec. It is the specific energy of the deuterium-tritium reaction.

There may be improvements in the theoretical design of a fusion-engine system, but the voyage length depends, of course, on the amount of fuel carried. A large fuel load could cut trip times given by Matloff in half.

A D+T engine uses massive quantities of hydrogen or deuterium. The fuel could be obtained from comets, the ice-covered surface of Jupiter and Saturn's moons, or asteroids. The lithium needed to breed tritium could come from an asteroid or from the moon. The total amount of lithium on the moon, based on some averages for lunar rocks, would be sufficient to launch a hoard of starships. If the asteroids have compositions similar to that of the moon (or if asteroids can be found that are similar) an asteroid about 4 miles in diameter could supply all the lithium necessary to fuel a D+T reactor for an interstellar flight. A D+T reactor uses lithium-6, which has an abundance of about 6-7 percent in lithium; it would take a massive amount of lithium to provide the Li^6, but there are plenty of asteroids available.

If all of the fuel for the starship *Agamemnon* were available in the icy satellites of the outer solar system and from the asteroid belt, gathering it would be, in itself, a monumental undertaking that might last as long as the construction of the ship. It is possible that other fusion reactions might eventually power the starship, such as $D+He^3$, but the D+T is more likely. Helium-3, if the former were chosen, could be obtained from the atmospheres of Jupiter or Saturn and the deuterium from rocky satellites, icy bodies, comets, or asteroids with high ice concentrations.

The idea of securing all of the fuel from the resources of the solar system is relatively new. Only recently was it discovered that the moons of Jupiter were partially ice covered and that comets might also be composed of large quantities of ice. The discoveries of the *Pioneer* and *Voyager* unmanned spacecraft and studies of comets have revealed that the outer solar system may once have had large resources of water that now remain as deposits on the surfaces of bodies there. From this it follows that there may be starship fuel in relative abundance for the builders of *Agamemnon.*

If fuel for a fusion starship engine can be produced in the next century, some of the parameters of *Agamem-*

non's voyage fall into place. If the ship carries 165,000 tonnes of helium-3 and 55,000 tonnes of deuterium (a "tonne" is commonly used in engineering problems for mass; it is a metric unit equal to 1000 kilograms), it would have a little over $2^1/_2$ times as much fuel for acceleration as the *Daedalus*. If it had an engine with the same general characteristics, *Agamemnon* could reach Alpha Centauri in about 200 years. Tau Ceti would take about half a millennium.

To reduce trip times still further or to make more distant stars accessible in the same time, there are several procedures available. The first is the use of electrostatic drag for deceleration. In effect, the electrostatic drag system is the reverse of the interstellar-ramjet. The starship would use a large net negative charge on surfaces about a kilometer in diameter. The net negative charge could be obtained by a direct interaction with the interstellar matter around the starship or could be produced by generators aboard.

Electrons would be deflected from the negative charge and protons would be attracted. This would result in a proton space charge controlled by a solenoid as large as 15 miles in effective diameter. Other devices would use charged grids to regulate the system, and a grid aft of the starship would be positively charged. The effect of this electrostatic shield—and there are several kinds possible for application to a starship—is interstellar braking.

The idea is attractive, but it presupposes advances that may not be part of future technology, although the concept exists on paper. Also, the method is not extremely efficient: a surface thousands of miles in diameter, if it could be deployed from the starship, would take 100 years to decelerate the ship from $0.02c$ to $0.001c$. Nevertheless, it would save fuel that could be used for acceleration and result in shorter voyages or in access to more-distant stars.

Another technique for saving fuel or improving acceleration at the start of the trip and decelerating the starship at the end is to use gravity assist. Using the gravity of Jupiter and Saturn, and perhaps of planets of a similar size orbiting the target star, *Agamemnon* could reach or decelerate from 0.001*c* without expending much fuel. As an alternative to carrying millions of tons of lunar slag as shielding, as proposed earlier, the starship could be double-hulled. Fuel would be carried between the hulls, which would eliminate an enormous mass that would otherwise require fuel to haul it up to cruise speed and down to target-system velocity. The fuel between the hulls would act as shielding for the starship.

The principles governing acceleration of a mass and the fuel required to do so are the same for a starship or an automobile. A 2000-pound car will use less fuel than a 4000-pound car, assuming the engines are about the same in efficiency and the aerodynamics are equal. The other side of the coin is that if the cars have the same horsepower, the lighter one will out-accelerate the heavier one by a substantial margin. The problem with *Agamemnon* is the drastic trade-off of great mass against acceleration. We know of no way to avoid the enormous mass needed to supply a crew of humans for hundreds of years. And we know of no way to avoid the environmental mass and still get the crew to a nearby star alive. It would be easy to theorize a cold-sleep starship with a small crew. One of those with the fuel load half that of *Agamemnon* would get to one of the closer stars within half a century or less.

But we don't know how to perfect cryogenic hibernation, and there are no indications we will know how to do so in the future. The breakthrough might be only a century away and it might be a thousand years. The Greeks knew about steam power—they used it in a toy and in a device which looked like an ordinary "Y"-arm lawn sprinkler—but it was almost 2000 years before the steam engine came into general use. The Egyptians had the basis of electric batteries and small devices which we think were used to produce electricity—but it was two millennia before electric lights illuminated the streets of our cities. The history of science and technology is filled with examples of a similar nature, and biology is no different.

Agamemnon will, at least, have at its disposal technologies which are logical extensions of what is happening now. That it will be incredibly slow is a problem, but for the moment, it is an insurmountable one. D+T fusion engines have a theoretical limit, mankind can survive space only in a closed and complex environment, as far as we know, and the stars remain as distant as ever. We can remove only so much extra mass from the starship; we can carry only so much fuel; and we are limited by our own humanity.

When the ship is completed and the propulsion section attached to the great cylinder, a preliminary crew will be drawn from the space habitats. They will join a trial crew of construction workers, technicians, and managers who will start the ship on a year-long checkout of each major division: navigation, propulsion, communication, and planetary science. The environmental, habitat, and computer complexes will have been thoroughly tested in the long decades during which the ship was being built.

Satisfied that the checkout is as complete as possible from orbit, *Agamemnon* will be given a shakedown cruise in the old-fashioned naval tradition. The shakedown cruise may consume half a decade or more; nothing as complex as the starship will ever have been built, and nothing in the history of mankind must be as reliable over centuries as the systems aboard.

The scene around the starship may be reminiscent of the departure of a great ocean liner in the days before commercial jets. Space tugs and construction vehicles of all descriptions and small craft from the space colonies will be gathered nearby. The casual and the curious, the important and not so important, the technicians and the crew will be assembled. The great starship will lie in orbit, spinning visibly. The smooth surface of the outer hull will be interrupted here and there by the dishes and shielded domes of the observatory.

The trial captain of the world's first starship turns a simple key about the size of a stubby pencil, and far away, at the other end of the *Agamemnon,* deuterium and tritium are injected into a thrust chamber. Passing the fusion point, the fuel is struck isotropically by a multipath megajoule laser for an infinitesimally tiny interval; the laser is focused on the fuel by optical mirrors. Ablatively imploded, a fraction of the fuel mass becomes the power of the stars. Inside the fusion fireball, two-thirds of the fusion energy is processed into motion of charged particles.

The stream of charged particles is collimated into an exhaust by forward and aft magnetic mirror coils. *Agamemnon* has ignition. Slowly, only a mile or so at first, the starship moves away from the orbit in which it has remained since its construction was begun. The time: probably the beginning of the twenty-second century, 140 years since mankind first landed on the moon, nearly 300 since the distance to the nearest star was measured; more than half a millennium since Galileo pointed his tiny "optik tube" toward the heavens and saw there the stars like dust.

The thrust of the starship is minute; the central sea has not risen far above its normal line. Ghosts of men and ghosts of dreams hover out in the orbit where it will pass near earth. Old Tsiolkovsky, the ever-present Russian dreamer, is there. So is Robert Goddard, the shy, contemplative American with his thoughts of the ultimate human migration. There are others: Hermann Oberth, Wernher von Braun, Eugene Sanger, Yuri Gagarin; the spirits of the three from *Apollo 204:* Grissom, White, Chaffee.

Near the earth's orbit rests the history of mankind's dreams of space. The wheel-like space stations of Noordung and von Braun, no longer in use; an astronaut's glove from one of the *Gemini* flights, still in orbit, tumbling from its original shove centuries before. The world's first solar-power satellite remains in place around the earth—in an orbit in which it will circle endlessly for another million years.

The high Clarke orbits—the geosynchronous ellipses about which Arthur C. Clarke wrote long before the Age of Space began—are filled with communication and television relays, giant antennas for contacting distant probes and the settlement on Mars, the scientific stations near Jupiter and Saturn; there is a great complex for electronic mail delivery to home terminals, new translantic cables of the future. There are bases on the moon, a giant observatory prying at the secrets of the universe; mining excavations and raw materials processing plants stand in the ancient soil of the lunar seas. The space colonies, holding a population of millions of the human race, move in their orbits near the moon and elsewhere.

The starship moves in toward the sun, past the orbit of Venus. On the hot surface of Venus great pilot engineering projects have been placed to change the atmosphere. Someday it may be another home for mankind, perhaps by the time the starship reaches its destination. It travels on past Mercury, and its vast cratered surface where scientists study the sun from up close, and mining operations carve the surface in explorations for new mineral deposits.

Swinging around the sun, the ship picks up a boost from the solar gravity and heads back out into the solar system. Still checking and testing systems and making thousands of adjustments and repairs, the crew guides the starship toward Mars. On the day *Agamemnon* passes the Red Planet, shadows of the moons of Barsoom play over the terrain; it is −20F on a late summer afternoon at the settlement near Solis Lacus. Men and women sitting in rooms on the only town on Mars run electronics tests for the starship. The weak sun glints off the ice and water surface of the single "sea" on the planet, near the settlement. On this world, so much a part of mankind's dreams of the conquest of space, the volcanoes of the northeast sector of Tharsis Ridge stand as silent as they did when they became inactive a million years before the penultimate ancestor of Man scratched in the wet sand by some forgotten seashore.

At a high inclination the starship passes over the asteroid belt, avoiding the bulk of the debris. Planetary-detection instruments are used on the bodies to test how well the subsystem will perform in its search for faint objects in a new solar system. On the rocks in the ring-shaped gap separating the Red Planet from the Giant one, miners run machines that chew and carve, digest and evaluate; mass-drivers send great bulky streams of raw materials to Mars, to earth, and the space factories.

The starship moves past Jupiter and the boiling hostility of its atmosphere, past the strange moons which are more like planets. The industry of the solar system is sifting the satellites of Jupiter in the search for raw materials and mining the miniature planets and the larger bodies. The system around the big planet is perfect for testing radio science, infrared radiometry, magnetic fields, plasma science, low-energy charged particle, and plasma wave detection equipment.

Hardly a century and a half earlier, the large moons that Galileo discovered in 1610 were first photographed by an unmanned spacecraft. The volcanoes of Io still throw gases into space, and the ashes rain down on a sterile plain of white sulphur dioxide snow. Europa is there, an infinite scrimshaw of dark lines, a surface of ice hummocks. Ganymede is visible, with its great flat stretches of frozen mud and ice. There is Callisto, a multilayered surface of craters—a world as dead as the earth's moon.

At Saturn the starship tests its instruments on Titan. Somewhere beneath the swirling and thick atmosphere of the moon, six men are toiling up a steep slope with geological instruments, their transporter floundering in the grooved surface covered with shale.

One hundred million miles from the orbit of Neptune, the starship starts back toward the sun in an enormous curving path. Somewhere far out beyond Neptune, the old unmanned spacecraft *Voyager*, which pointed its cameras at Neptune in 1989, is still spinning its way into dark space. If the nuclear-power unit is still functioning and any of the radio equipment has not completely ceased to exist, the starship might listen in to one channel of the ancient spacecraft, still telling its story of a final voyage.

The shakedown cruise will be over when *Agamemnon* settles slowly toward a parking orbit out beyond the moon. The tests and repairs have been logged; the failures have been noted. There will be more fitting and checking out. Then the final crew will come aboard and fill their cubicles and apartments, arrange their lives, man their stations. The voyage will begin.

13 THE COATTAILS OF GOD

"The foxes have holes, and the birds of the air have nests; but the Son of Man hath not where to lay his head."
MATTHEW 8:20

Acceleration aboard *Agamemnon* will be leisurely. Gagarin's *Vostok* had 500 times the starship's acceleration during the first few minutes of lift-off. Space shuttle, which is designed to be easy on astronauts, has an initial acceleration 150 times more. The virtue of *Agamemnon* is that it can maintain the acceleration for a very long time. At the end of the boost period, the cruise velocity will be a small fraction of the speed of light.

At the cruise velocity which the starship will attain, the relativistic effects of starflight are small or nonexistent. When built, it is unlikely to exceed 10 psol, and only at velocities greater than that are relativistic effects particularly noticeable.

Apparent shifts of stellar positions caused by the forward motion of *Agamemnon* and apparent shifting of stellar wavelengths will be minimal and compensated for by computer in the navigation complex. As for the crew, the apparent color changes of stars associated with wavelength shifting will hardly be noticeable, even from whatever serves as the bridge of the starship. The relativistic time effects will be minimal: at the distance of Alpha Centauri, the ship's clocks will register a few months "earlier" than the equivalent clocks on earth. Nevertheless, it may be the largest relativistic time effect mankind will have seen: the fastest spacecraft, *Pioneer 10* had a "clock" only a few millionths of a second faster than the equivalent on earth.

What is the destination? Will techniques such as aperture-masking, electronic sifting of images, the use of a telescope in space, and speckle interferometry* have eliminated most of the nearby stars as serious candidates for man's planets? Will one or more starprobes have reported back to earth that there are green and blue and wonderful worlds rather like the earth circling Alpha Centauri, Barnard's Star, or one of the other close possibilities? From all the statistics and educated guesses so far compiled on habitable planets nearby, the best destination for *Agamemnon* might be Epsilon Eridani.

Epsilon Eridani lies, as might be assumed, in the con-

*Speckle interferometry uses sequential high-speed images, which are correlated in a way that compensates for the distortion caused by the earth's atmosphere in earth-based telescopes. It is a technique used to resolve very close binary stars—or nonstellar companions such as planets, perhaps. In 1980, the technique indicated the asteroid Pallas may have a satellite 175 miles in diameter. The application to detecting planets around other stars is obvious.

stellation Eridanus (the river). Eridanus is a long constellation extending from the celestial equator far into the depths of the southern sky. The brightest stars of Eridanus are far enough south that they do not rise at the latitude of New York, although they are easily seen from Mexico City. Alpha Eridani is a star known since the time of the earliest Arab astronomers. Also called *Achernar,* it is one of the brightest stars in the sky. The constellation is relatively undistinguished except for a well-known spiral galaxy called NGC 1300.

The two principal attractions in Eridanus are Epsilon and Omicron, the latter also known as 40 Eridani. Omicron Eridani is almost 16 light years away and it is a multiple system with a main star like the sun, a second star that is a white dwarf, and a very faint third star that is a red dwarf. Epsilon Eridani has remained a candidate for habitable planets through most of the elimination rounds in the statistical analysis so far made. It is 10.69 light years from the sun, more than twice as far away as the nearest star.

The mass of Epsilon Eridani is estimated at 0.80 solar masses, meaning it is, of course, smaller than the sun; it could be about the same size, but the mass estimate would then indicate a lesser density, and this is not likely. The star is located 10° south of the celestial equator; it can be seen from most regions of the earth except the poles. It is a single star with a visual magnitude of 4.2 and thus can be seen with the naked eye or with binoculars, although it is not particularly bright.

The estimates of chances for a planet habitable for mankind around Epsilon Eridani have varied from 1 in 30 to much better. The principal reason for the variation in estimates is that the star is a K2 spectral class, cooler than the sun. Coupled with the smaller mass, it means that the zone of habitability around Epsilon Eridani is small and a planet would have to lie very close to the zone to be habitable. The chances for an exact formation at exactly that distance are about the same as the chances that the earth formed in the way it did. That is, it's likely that it could happen because it happened in our own solar system. This probability is, once again, based on the assumption that the earth is not in any way unique. In 1950, the coordinates for Epsilon Eridani were: 3hr 31m R.A.; declination $-9°$ 38′.

The star was listened to in the spring of 1960 during part of Project Ozma. At the time, it caused quite a sensation. The astronomers of Ozma were using the 85-foot dish of the radio observatory at Green Bank, West Virginia. On April 8, 1960, they turned the antenna on Epsilon Eridani. Shortly after the dish was locked on the star, the recorder trace needle went right off the scale. The trace showed a series of high-speed pulses, very uniformly spaced at about 80 pulses a second. Nothing anyone could imagine could have produced the pulses naturally, and pandemonium took over the control room. Most of those listening to the pulses coming in on the speakers and watching the trace were sure that an extraterrestrial civilization had been discovered. Somewhat mysteriously, the radio emission ceased after five minutes. After the first rush of excitement, the astronomers were cautious. All of the equipment was checked out but nothing unusual was happening in the electronics.

Two weeks later, the same signal was again heard, but this time the astronomers had time to steer the radio telescope antenna away from Epsilon Eridani. The signals remained loud and clear; the needle trace showed a steady pulse pattern. As it later turned out, the strange pulses came from military radar-countermeasure equipment aboard aircraft. In the three months following, no further signals seemed to originate from Epsilon Eridani. The experiment proved several points about extraterrestrial communication: the 21-cm line of hydrogen, which was the wavelength listened to, might not be the most suitable; and, in any event, it was subject to interference from aircraft and possibly a half dozen other sources not associated with alien civilizations.

That no signals came from Epsilon Eridani proved nothing. If a culture on a planet around the star had listened in on the sun-earth system in the same manner with the same equipment, it also would have detected nothing. Mankind is not sending out any transmissions on 21-cm wavelengths.

Whether the crew of *Agamemnon* will find a civilization at Epsilon Eridani is doubtful—but the likelihood is that they will find planets. All our statistical models show that slow-rotating, average-aged stars in the galaxy probably have one or more planets of some kind. The target star, being a single star of the right rotation and age, will have a small system of planets. What the crew will hope their descendents will find is a planet which has formed at about the same distance from Epsilon Eridani as Venus has from the sun. Such a planet would have a good chance to follow an evolutionary track sufficiently like that of earth to make it suitable for mankind.

The crew will have a long wait for the answer. At the stately crawl of *Agamemnon,* the starship would now be arriving at Tau Ceti if it had been launched during the Renaissance. Had it been sent toward Alpha Centauri during the American Revolution it would still be on the way—arrival scheduled for A.D. 2005. Though such long voyages have their disadvantages, there are positive aspects. The ship is a completely closed ecology designed to last thousands of years. As far as the people on the ship are concerned, there is no more advantage in going to the nearest star than there is in going to one which might have a much better chance of having a habitable planet—even if the better possibility is much farther away.

For those aboard *Agamemnon,* life will be the *interior;* there will be only the ship. History will be a distant dream and the earth a distant memory kept alive by images in the learning rooms and an occasional message from the communications link. After a few generations aboard, the time will be long past when every message from earth will be posted on the equivalent of a community bulletin board; and those that are posted will probably not elicit a great response. Would we, in the twentieth century, be interested in the customs, clothing, politics, of Italian city-states in 1515 on a daily basis? Perhaps so, if we were historians or the keepers of some eternal flame, but most of us would be far more interested in what we'd be having for dinner or what was on stage in village-3. Still, the world of the late Renaissance is probably less familiar to most of us today than the earth and its culture will be to the twelfth or thirteenth generation aboard the starship. They will at least have the advantage of communication, which we are denied.

The isolation from the rest of humanity may not be one of the serious problems of the voyage. Human populations have existed in isolation on earth for more than several generations. The first Europeans to enter the vast reaches of the Pacific discovered, much to their amazement, that living on a number of small mountainous islands and flat, sun-bleached coral reefs was a population of tall, handsome people who walked out to meet them, accompanied by dogs, and familiar birds, and pigs. The islanders spoke a common language, but it was a language unknown in the rest of the inhabited world—Polynesian.

Writing was not known among the Polynesians, with the exception of wooden tablets faithfully kept on Easter Island. Not one man or woman on any of the islands in all of Polynesia could read the strange, incomprehensible hieroglyphs on the tablets. The inhabitants of the islands did, however, have schools, and history was their most important subject. For them, history was the same as religion; they knew their heritage all the way back to Tiki, who was, they said, son of the sun.

The peoples of the islands in the Pacific had apparently arrived about A.D. 500 and had remained culturally isolated until about 1100, when a new wave of migrations reached the islands. From 1100 until their discovery by the Europeans, they were completely isolated and completely self-sufficient from the rest of the world and its civilizations. Without communication with the rest of the world, without benefit of science or technology, without the overwhelming blanket of man's great religions, they loved, lived, and died in peace, in isolation, far longer than the people aboard *Agamemnon* will be expected to endure. While they retained without substantial progress the basic culture which their earliest emigrants had possessed, they nevertheless were a stable and relatively healthy group of humans.

The crew of *Agamemnon* might originally number 10,000. Of that number, about half would be within the span of their normal working years, and a quarter would be of school or college age. Of the working group of the crew, 500 would be employed as teachers, 500 as attendents of *Agamemnon*'s farm, and another 500 would perform the tasks required by the computer complex, maintenance, and navigation. About the same number would be the shopkeepers, small businessmen, librarians, nurses, doctors, and restaurateurs. Perhaps a thousand of the crew would be engaged in small manufacturing projects designed to replace perishable items.

The remainder of the crew of working age would be scientists of one brand or another: astronomers, physicists, mathematicians; a small number would be the "line officers" directly responsible for the performance of the starship, and the chaplains. The community would probably not be unlike one of the isolated atolls of the Pacific—relatively homogenous, stable, and diversified enough to survive generations without constant updating from the outside.

Will the builders or crew create a truly strange and unusual architecture? Will the long voyage so change humanity that the interior of the ship would be unrecognizable to us in the twentieth century? Drastic changes are unlikely, because humanity has common needs. We still have two arms, two legs; we will remain about the same height. We still need to brush our teeth, comb our hair, place part of our anatomy in something resembling a chair, walk through something resembling a doorway. To paraphrase Dr. Freud, a cigar will probably resemble a cigar.

Imagine an officer of a wooden man-of-war in the time of George Washington who was suddenly (and inexplicably) transported to the wardroom of a nuclear-powered, missile-equipped supercruiser patrolling off the coast of North Africa in the Mediterranean. After the initial shock, what would he recognize on this ship built two hundred years in his future? A plate would be a plate. A fork, a fork. Men in white uniforms bringing food in stainless-steel containers would be recognizable. Water glasses would be water glasses. A salt shaker would be exactly what you would expect. Salt shakers have changed very little in general since mankind began putting salt in containers. Whether it is an abalone shell from which salt is pinched with fingers or a "modern" glass and stainless-steel affair, the use is obvious to a hungry human being.

Aboard the nuclear cruiser would be chairs for sitting down to table, and napkins. The food would be considerably more interesting than salt pork, but quite recognizable as an evening meal. In short, our theoretical, transported Revolutionary War sailor would be able to sit down and catch up on two hundred years of history while having very little trouble with the environment.

He would be instantly aware of the purpose of a bunk, and a wash stand would require very little puzzling over. The "head" in officer's country might be intimidating, but the use would be obvious. It is not in the area of human needs that mankind's technology has made vast changes; rather, it has been in medicine, and communication, transport, and, unfortunately, war. Our sailor from the past would understand the use for but would hardly recognize a sick bay aboard a modern ship. The engine room would be a complete mystery. (The lack of sails, once he got on deck, would seem a bit odd, too. He might conclude the ship was manned by devils and driven by magic. So would the weapons of war, missiles that could reach, in an instantaneous twist of a key, based on a command from thousands of miles away, a target in parts of the world unknown to him.

But if he kept his sanity, within a few days he would be roaming the ship with relative confidence. He would not need to have a blockade explained, nor, once he was used to the method, the source and reason for orders. The purpose of a military ship has been much the same for over two thousand years. Only the ability to wage war and the conveniences aboard have changed drastically since the days of the Trireme.

In somewhat the same way, *Agamemnon* may not appear that different to our twentieth-century eyes. For us, too, the "engine room" might be basically a mystery, and so might the "bridge," but the living areas would be quite familiar. Like our sailor from the past, we might find the sick bay a little strange.

Taking a person's temperature with a thermometer is an unhandy necessity of our time. So is plunging a hollow needle in a vein to draw blood for laboratory examination. Diagnosis aboard *Agamemnon* will probably be practiced by remote scanners with a bedside manner managed by an "intelligent" computer. The sick bay may have built-in panels above the beds to monitor a patient and read out details of the medical parameters. Some of these types of machines are under development in our own time. While a machine may never replace the doctor, the "GP" aboard the starship may be a blend of wires and circuits that admirably duplicates the function of an M.D.

The laundry facilities may also be recognizable, but

instead of washing and ironing or dry-cleaning, the machines may simply reprocess the garments. The dirty clothes would go in one end, be reduced to the original chemical fibers, dirt would be removed, and the garment would issue out the other end of the machine, not only clean but new.

The real change aboard the starship may be social and cultural. In this arena, people have always varied immensely and will be no different out among the stars. What will constitute "good" behavior or proper etiquette may differ greatly from that today. Aside from the obvious ways in which things might change—acceptable language, courting patterns, folkways, forms of address, and attention (or lack of attention) to personal privacy—there is the matter of crime.

What would constitute a crime aboard the starship? Would it be the same sort of antisocial act we have known for centuries, even millennia, or will there be strange and unexpected variations? Murder, rape, assault—crimes which we have punished throughout history—will probably be just as punishable aboard *Agamemnon.* Barratry, which is in maritime law a fraudulent breach of duty by the master or crew with the intent to harm the owner or cargo, would, aboard the starship, approach the level of a crime against humanity. There will also be "social crimes." Wastage would be the most indefensible with the limited amount of irreplaceable material, and perhaps the most often punished offense.

Theft may be almost unknown. The two principal motivations for theft are different aboard from what they are on earth. Theft involves profit, which in turn involves resale. The community aboard the starship is too small to make the occupation of "fence" very desirable. Although it is true that petty crime and theft occur in small communities on earth, as well as in large cities, it is also possible, on earth, to go elsewhere with the goods.

Aboard *Agamemnon,* something stolen from village-A would probably be recognized in village-B. A thief would not be able to hide easily aboard ship—there are only so many places to go.

While death has been the ultimate punishment for crimes in most civilized countries of earth, it might not be the ultimate aboard ship. Perhaps the ultimate punishment would be removal from the voyage. The prospect of hanging or lethal injection or some other state-controlled capital punishment has not been much of an effective deterrent in our history. But there are more kinds of death than one. How about being thrust out a remote airlock of a starship traveling at a rate of millions of miles per hour billions of miles from nowhere?

When the last instruments are aboard, when the last of the crew has been assembled, the great starship *Agamemnon* will lie in orbit, waiting for the Beginning, as the people aboard will call it in generations to come. The scene at the great ship's departure on the shakedown cruise was only a foretaste of what will happen when it finally departs the solar system for the planets of Epsilon Eridani.

Once again, the captain will insert his key in the control room and the ship will slowly move away from its parking orbit. It will be guided along a spiral trajectory onto a flight path which will eventually exit from the neighborhood of the sun. Noiselessly in space, with no thunder of rockets, *Agamemnon* will drive away from the sun along a narrow channel defined by the navigation computers. Centuries down that channel, the planets of Epsilon Eridani will come into view. In between, more than a dozen long-lived generations will work, dream, and die aboard.

If gravity-assist is elected as a method of ejecting the starship from the solar system, it will head outward toward Saturn. After months in space, it will come back toward the inner planets in a great rebound toward Jupiter. As the giant planet is approached, it will begin a powered close flyby, a peri-Jove maneuver. From the bridge, the swirling atmosphere of Jupiter will appear frighteningly close. The red clouds, the top of the Great Red Spot, and Jupiter's white ammonia-crystal clouds will be easily visible.

The complex Saturn-Jupiter maneuver will leave the starship with a high-velocity solar-hyperbolic flight trajectory. It will travel toward the inner planets, gathering speed. As the great starship drives toward the sun, the boost will be increased to 0.1 g for three days before the closest approach in the perihelion maneuver. The ship will be drawn toward the fiery atmosphere of the sun, assisted by its own acceleration. At the perihelion maneuver, with the vast boiling globe filling the space around the ship with particles and heat, *Agamemnon* will make its closest approach, closer than mankind will have ever gone to the star which it has known as the sun throughout all of its history.

The velocity at the completion of the perihelion maneuver will be 372 miles per second, 1,339,200 miles per hour. The navigation computers will lock on Epsilon Eridani and the voyage will begin. At first, the starship will use the main engines to raise the solar-system escape velocity—1/100th gravity acceleration will be added to the velocity acquired by the gravity assists of Saturn, Jupiter, and the sun. For a year, the crew will be busy with final adjustments and navigation checks. Using star positions referenced to distant quasars, they will have aligned *Agamemnon* perfectly with its flight path. The Captain will order 1/50th gravity for the main boost into interstellar space.

Once in the main-boost phase, the routine of the ship will begin. A steadily increasing stream of information about interstellar space will be transmitted with ever-increasing lag back to earth. The lengthening baseline between the ship and the solar system will be used for measurements of the galaxy and a series of complex scientific explorations in physics and astronomy. Slowly the boost will build up until cruise velocity is reached.

Finally the engines will shut down and the ship will go forward into interstellar space at its final velocity, being slightly and infinitesimally retarded by impacts with dust and tiny particles. The original crew will die and be replaced by men and women to whom earth is only a recorded memory, men and women to whom only the voyage is real.

Several light years into the voyage, the ship will register the lessening of solar gravitation. At the point where it is only a tiny fraction of the pull near the sun, the region between the stars, true interstellar space, will have been reached. While there is no cosmic signpost which says "You have reached interstellar space," the crew of the ship will suitably mark the event as a reminder to those who live aboard the great cylinder that it has a purpose beyond the day-to-day existence in the cities and villages of the interior.

At midpoint in the mission, the ship will be turned so that the engines are facing toward Epsilon Eridani. If the technology was available at the time *Agamemnon* was constructed, the ship will use an electrostatic drag brake deployed during the generation immediately following the midpoint turn-around. The initial effect of the electrostatic drag will be tiny, but over generations, the cruise velocity of the starship will drop slowly toward the point at which the engines will be restarted and the main deceleration phase will begin.

Before the engines are restarted, with the ship still slowing by electrostatic drag on the interstellar medium, the planetary-search optical and radio telescopes will go into operation, aimed at the target star. At 5.07 light years from Epsilon Eridani, 5.62 light years from earth, Jupiter-size planets will be separated from the light of the star. The optical telescopes, using the special instruments attached, will show the 24th-magnitude object as a gas-giant planet.

At 2.5 light years from Epsilon Eridani, very faint companions of the star will be visible or detected from the observatory—earth-size planets. At this point in the voyage, nothing can be determined about the worlds except their existance; but for the crew, it will be another sign that the voyage will eventually end. Someday, their descendants will arrive at the new solar system. While the discovery of Jupiter-type planets will be exciting to the crew (although they may have been known previously from earth-based observation or starprobe investigation while the ship was being built), the discovery of earth-size bodies will be a major milestone. From planetary detection, there will be a mounting excitement aboard.

Engine restart will be accomplished and 1/50-g deceleration will take generations. Stopping the ship will require almost as much thrust for almost as long as reaching cruise velocity did—not quite as much, and not quite as long, because the ship has less mass because of the fuel burned in the lengthy voyage out. Finally, the ship will encounter the gravitational sphere of Epsilon Eridani. At 0.4 light years, the Jupiter-size planets can be seen in the telescopes without special instruments. At 0.2 light years, the earth-size planets are also visible without special equipment.

As *Agamemnon* goes in toward the new star, details on the large planets will be visible. Details and characteristics of the planets like earth, if there are any, will become clear as the starship approaches 20 billion miles from the Epsilon Eridani system. From that point on, the crew will be deeply involved in exploration of the new worlds by telescope, radio astronomy, and all the techniques which mankind may have developed for remote sensing on distant objects.

At about 100 billion miles, *Agamemnon* will be on an approach orbit to Epsilon Eridani. The planetary-search computers, assisted by some of the crew, will energize, test, and release a series of planetary probes. The probes will be computer-controlled and highly sophisticated versions of the kind of unmanned spacecraft which explored the solar system in our own time. They will use radiometry, photopolarimetry, direct-imaging cameras, spectroscopy, and other techniques to relay a steady stream of information. At the same time, the vast computer complex aboard *Agamemnon* will send digests of the data to earth, where it will arrive in about ten years. It will be the first information from another star system sent by a branch of the human race. It will arrive centuries after the *Agamemnon* launch.

Depending on the goals and equipment, the planetary probes will be sent on separate missions. Some will be targeted for the giant planets of the Epsilon Eridani system and will investigate the environment around the planets and the satellites. Some will go to the rocky, terrestrial bodies. The probes for the inner planets will be orbiters and landers. The main bus will remain in orbit over the planet and complete a mapping sequence. The orbiters will investigate the chemical, mineral, and land distribution on the worlds. Computers aboard the starship will reduce the massive information pouring in from the probes and compare it with any information which was previously obtained from earth-based telescopes or earlier starprobes.

Or perhaps none of these probes will be necessary: the whole idea of unmanned planetary craft like our *Voyager* or *Viking* may be completely obsolete. Maybe the *Agamemnon* will use some esoteric variation of a long-range device based on the gravitational lens which will reveal all the crew wants to know about the new planets from several light years away.

Whatever the planetary distribution around Epsilon

Eridani, the starship will be targeted to enter the main body of the solar system and proceed to a parking orbit around the new star. From this vantage point, the *Agamemnon* can spend as much time as necessary to investigate the entire system for habitable planets and potential resources. Should they choose to do so, the crew could spend a hundred years studying the planets before they try a landing on one of them.

What if the ship detects artificial radio emissions from one of the planets? No instructions would be obtainable in time from earth, with a communication lag of nearly 22 years, for a round-trip message. The crew members would be on their own. They would notify earth of the situation and then decide for themselves what to do. The crew might attempt communication by similar frequency radio waves. Common physical constants such as the speed of light would be transmitted in an effort to establish the starship as an artificial transmitter and "civilization." What might result from a two-way communication would probably both fascinate and terrify both cultures. For the other culture, the situation would be analogous to what would happen if an alien starship showed up in geosynchronous orbit next to one of our television satellites and interrupted the Saturday Night Movie.

If the evidence of a civilization on one of the planets came in from early observation the starship might choose to establish a distant parking orbit and evaluate the situation. There is just as much reason to be paranoid as not, and the branch of the human race aboard *Agamemnon* might decide to proceed very slowly into the Epsilon Eridani system.

If the planets around Epsilon Eridani are not inhabited—and given the statistics we now use, it would seem unlikely—the starship will sit in its parking orbit until the crew are ready to try a landing. Whether they actually take a trip down to the surface of one or more of the new worlds depends on the conditions there and how adven-

After the Agamemnon is in orbit in a new solar system, the crew will launch shuttle-craft toward the surface of one of the planets which is most suitable for human habitation.

turous the people of *Agamemnon* are after generations aboard the closed ship. No one alive will have any direct knowledge of how to pilot a spacecraft or an atmospheric flight vehicle. Trying to do so from computer tapes might prove hazardous.

Given the proper extensive information, computers which will have been developed during the time of the starship's construction might be able to land a vehicle without a human pilot aboard. The computer on the Mars-Viking lander set the craft down quite gently on a planet whose exact surface conditions were imperfectly understood. It is also true that there was considerable luck involved—one of the Vikings narrowly missed a field of boulders, any one of which would have pierced the underneath of the craft and destroyed it during the landing. There was a reason that the *Apollo* LEM had pilots as well as computers: the last few minutes of a landing are hard to judge. Computer technology, however, by the time the *Agamemnon* is built may be so fast and intelligent that they could manage a planetary landing easily.

It is unlikely that the human race aboard the starship will long remain in a parking orbit. If there is a habitable world, at least some of the crew will make the trip down to the surface eventually. For them, stepping out on a planet will be a new and frightening experience. A horizon so distant that its curve is scarcely visible will be something the starship people will never have seen except in photographs lodged in computer storage. They will also never have felt rain, or seen a sunset, felt or heard the waves of an ocean, the rhythmic beat of the sea. They will never have seen a night with a moon, nor an eclipse, nor a thousand little things we have all taken for granted over all of our lifetimes.

No matter what the outcome, some of the crew will choose to live on one of the planets for a time. Our history is full of reports of the few who stayed in a new land to try and make a colony flourish. There will be those among the people of the starship who will make the gamble. They may even carry their technology down to the planet and create small towns and villages much like those they left in the artificial world still orbiting the strange sun as it rises and sets.

What if Epsilon Eridani has no planets where mankind can live? There are several possibilities. The ship could simply stay in orbit, and life would continue aboard as it had done for centuries. The crew could use the resources of the solar system to refuel. The icy satellites expected around a giant gas planet would serve, just as they served in the old solar system of mankind. *Agamemnon* could return to earth, though there would seem very little reason to do so.

It is more likely the crew would elect to continue the journey—onward to another, relatively close star. They would continue their primary mission of finding a habitable planet for mankind. Another star is close enough for a dual mission: Omicron Eridani, the strange triple system described earlier. Though the chances for a habitable planet in a binary or complex system are not now regarded as good, our theories may be revised by the time the first starship leaves our solar system.

Omicron Eridani is 6 light years beyond Epsilon and 11° line-of-sight separation in earth's sky. With a 30° turn angle at Epsilon Eridani, the second star could be reached at about 7 psol in a few more generations. Whatever happens in the constellation Eridanus when *Agamemnon* arrives, mankind will have fulfilled a magnificent dream and gone wrapped in its technology to the planets around another star. Nothing will ever be the same again.

14 YOU CAN'T GO HOME AGAIN

"To make available for life every place where life is possible. To make available all worlds as yet uninhabitable and all life purposeful."

HERMANN OBERTH, 1923

Centuries-long voyages may not be what mankind has in mind for starflight, eventually. We were not satisfied with Lindbergh's 30-plus-hour flight across the Atlantic in 1927. It only whetted our appetite so that in 50 years we had perfected the Concorde to span the distance to Europe in a few hours. Mankind's future scientists may find ways to perfect antimatter propulsion, and they may yet find a way to circumvent the pretty rules of Einstein and exceed the speed of light. Starships like *Agamemnon* may become obsolete, indeed.

An antimatter starship, if it becomes available at some point after fusion ships are common, will probably carry its fuel as a solid—frozen antihydrogen at a temperature only a few degrees above absolute zero. The atoms of the fuel would consist of antiprotons and positrons. The antihydrogen would mix with "normal" hydrogen at ignition. The antimatter would be maintained by electrostatic fields and protected from collision with ordinary stray matter in interstellar space by thick shields. If it could be carried safely, antimatter would last indefinitely—that is, it would remain stable for use over a very long period of time.

An antimatter-powered *Agamemnon* would be able to travel to the stars in much less time than our fusion ship to Epsilon Eridani. If the M-AM engine is developed, a small, highly specialized *Agamemnon*-type vessel with a crew just large enough to constitute a stable society could make the trip to Alpha Centauri in only a few years. Barnard's Star would take perhaps a decade; Epsilon Eridani might take 15 years or so. The development of such ships would begin the colonization of this part of the galaxy, because the trips, though long, would be within the lifetime of a human being and the result would be a slow but perhaps inexorable emigration to the planets of other stars.

What if the speed of light could be exceeded? "The mechanism of rocket acceleration," Dandridge Cole once wrote, "is fundamentally different in character from any of the phenonema which have been used to check relativity." While there are those who would argue with Cole on that, the statement, as it stands, is essentially true. It may happen that our descendants will be able to power starships with antimatter up to *c*—for the first time we would be able to test the "barrier" of the speed of light directly with an experimental starship.

And if such a ship did exceed the speed of light by some fancy mathematics or great witchery, the human migration to the stars would be immediate and expanding. If we fall back once more on statistics, we find that even with ships capable of $0.1c$, which is within the sort of technology that would build the *Agamemnon,* the Milky Way could be colonized in as little as 1 million years. This, of course, assumes that more than one multigeneration starship is built.

Other ideas of the density of stellar populations and the possible effects of emigration (using as models the Europe-to-America migrations of the eighteenth century) have put a figure of 10 million years on how soon the galaxy could be colonized by $0.1c$ ships. While this is an immense interval in the history of humanity, it is a tiny interval in the lifetime of the Milky Way and an infinitely tiny interval in the existence of the universe.

Antimatter ships would drastically cut the times estimated because of the ability to reach high fractions of c; if c could be exceeded, then mankind would spread outward in a fan-shaped emigration to the nearest stars within a few centuries. There would be outposts of man everywhere within this arm of the Milky Way.

The idea that ships can someday exceed c involves a bit of semantics. From the viewpoint of a starship crew, considering relativistic time effects, the ship *is* going faster than light at, say, $0.99c$. At that velocity, the craft would travel a light year in 0.14 years crew time. The apparent mean effective velocity, in that case, is seven times the speed of light. It is only from the viewpoint of earth that the trip would take a year at fractionally sub-light speed. What is generally meant in discussing faster-than-light ships is high multiples of c from the earth-based point of reference: starport New York to starport Tau Ceti II in a week, or something like that.

Imagine a far-distant future and its starship: the ovoid shape is designed for a crew of 10 with a cargo of 150 colonists in frozen sleep; along with their equipment the colonists will be opening up a small world around Arcturus, a K2 IIIp star 36 light years from earth. The ship departs from the starship base around the sun using antimatter power from fuel pods underneath the vessel. Near light speed is reached in a few weeks. Just under *c*, the crew set their control boards for multiple-light-speed maneuvering. They retire to their compartments while a slight twitch happens somewhere behind them in the engineering section. A tachyon drive has taken over and the ship is now streaking across the galaxy at 600*c*, relative to the earth. In less than a month, the ship, completely controlled by computers buried in the atomic structure of the hull, arrives at a point a few light years from Arcturus. The tachyon drive ceases and the crew go on into the interstellar space near the star at sub–light speed. The colonists are left on the planet around the star and the ship heads home.

While this is the "All ahead warp factor 5, Mr. Sulu," level of starflight, there is no reason to suppose that it may not be a part of mankind's future. None of us, not the most poetic, the most imaginative, the most technical, nor the most fey, would have, in the fifteenth century, been able to predict with any accuracy the world of the twentieth.

What will make human migration into this arm of the galaxy perhaps inevitable, if the propulsion and speed-of-light problems are finally solved, is what has been called "planetary engineering." There are only a handfull of habitable planets expected to be found circling the stars in this local area of the Milky Way. Mankind will, of course, eventually reach and probably colonize those. But out among the hundred nearest stars are as many as a thousand planets, possibly more. Most of these are not what we consider habitable planets for mankind as we understand solar-system formation now. But they could be *made* habitable. They could be forced onto an evolutionary track which would make them more earthlike. It may be possible for mankind to make even a very inhospitable planet habitable.

The beginnings of planetary engineering—making a world into a new earth by giant engineering techniques—will be tried out in our own solar system first. It has been suggested that Mars could be made into a habitable planet comparatively easily, since it is, in gross terms, not as much dissimilar from earth now as it is simply different. Mars is a planet which has followed a slightly different evolutionary track since the early days of the formation of the solar system.

If nuclear devices were exploded on Mars in the right place, the volcanoes would come to life again. The devices could be delivered to orbit by unmanned spacecraft and dropped with pinpoint accuracy (a legacy of military interest in ballistic-missile targeting on earth). The volcanoes would begin out-gassing into the atmosphere, as Mount Saint Helens did on earth and as all volcanoes do all over the solar system. Most planetary atmospheres on the larger planets are thought to be a product of volcanic out-gassing. On Mars the result of this event would be a thicker atmospheric envelope with greater atmospheric pressure.

A somewhat gentler method would be to establish giant solar mirrors in Mars orbit. The mirrors, great, spidery structures developed from space industrialization in near-earth orbit, would focus sunlight on the Martian soil. Other mirrors would be focused on the Martian polar caps. The soil would heat up and release gases into the thin Martian air. The polar caps would melt, releasing water vapor, and carbon dioxide; the result would be a higher atmospheric pressure and a thicker atmosphere.

The combination of pressure and density of atmosphere on Mars would have a twofold effect: water vapor would eventually fall to the surface as rain (and there would be enough pressure for water to remain liquid, which is something which it cannot do now on Mars); and the denser air would shield the surface against the bath of ultraviolet radiation it receives from the sun now. The combination would also allow more heat to be trapped during the day, and this would begin a cycle of heating up the planet which would release more gases trapped in the soils and slowly melt more of the polar caps. Present subsurface ice would become liquid, which would moisten the Martian soil. Mars would go from being terribly cold and arid to simply cold.

Some preliminary studies have indicated that if mankind were willing, the technology of solar mirrors and vast changing of the atmosphere of Mars could be begun in this century, or certainly within the first quarter of the next. As with nearly every other extrapolation of space development in the future, the idea is dependent on the space shuttle's paving the way.

Once the atmosphere of Mars was changed in density, the constituents would undergo a radical change. The gases released from the soils of Mars and the polar caps would not necessarily make a breathable atmosphere by human standards, just a thicker one. Bacteria, specially developed on earth through rDNA techniques, would be seeded on Mars. These would change the atmosphere eventually to the point where it would be useful for mankind. Two bacteria already studied as possible planetary-engineering agents for Mars are *Azotobacter beijerinckii* and *Beijerinckia lacticogenes*.

What can work on Mars can work on any similar planet around a nearby star, which would double the possibilities for habitable planets for mankind in nearby interstellar space. If we could learn how to engineer a change on Venus which would result in its becoming more earthlike, this would triple the chances. If we know how to modify three of the planets in this solar system to sustain mankind, then any three other planets among the stars of size and heat similar to the sun's could also be modified.

Beyond modifying Venus, it has been suggested that even the moon could be given an atmosphere suitable for humans if solar mirrors were trained on the lunar soil for several decades, perhaps half a century. While the atmosphere would be slowly lost again to space by natural lunar mechanisms (the moon is too small, the gravity too low, to retain an atmosphere for more than a few thousand years), it could become a tourist world in the far future. And if we wanted, other moons of this or any other solar system could be made into small habitable worlds for mankind. Perhaps we may one day find the secret to gravity and be able to generate it without having to rely on such old-fashioned things as mass. If that were achieved, we would simply plant a gravity generator on the moon, bring the moon up to near earth gravity, and start generating an atmosphere. The moon could hold an earthlike atmosphere for tens of thousands of years at 0.8 earth gravity.

We could even change the rotation of planets using planetary engineering. By sending asteroidal bodies into the moon, or into Venus (which has a slow rotation, too), at a sharp angle, we could effect a spin-up of the rotation. Eventually, the moon would have a "day" like the earth's, and an atmosphere, and plants and animals—exactly as man presumed it had in the sixteenth century.

With fast starships and planetary engineering, all of the planets available to us in the local star population could be colonized. Ships with crews of cyborgs or constructs could be sent to the more difficult planets; "normal" earth people could be sent to the planets like Mars and Venus which would be easier to change. Perhaps someday von Neumann devices will go first and prepare the worlds. Mankind will stay home until the new planets are ready for habitation. Let the robots do the work.

Maybe some far-future earthman will be at the controls of a strange spaceship which drives into the regions around a black hole and through a bridge to another universe. That we do not now think that such a bridge exists or can exist may not matter; we have been wrong about the universe before.

And what of *Agamemnon*, so long ago left in a parking orbit around Epsilon Eridani or Omicron? It will have established a colony and perhaps journeyed onward to yet another star. Eventually, faster starships may catch up with *Agamemnon*, whose position will still be known to earth from data streams received across the light years. Someday it may be turned into a museum, and people will wonder how anyone had the courage to climb aboard a giant coffin and go off adventuring among the stars. By then, some mad teakettle accelerated by burning bits of separate universes may have already whizzed past the memory of Einstein.

People of the galaxy will stand, silent and tiny, on the plains of the great cylinder and look in awe at the central sea. They will ask, as they grow older, if they are made of the same stuff as the crew of *Agamemnon*. They are.

"We are all in the gutter, but some of us are looking at the stars."

OSCAR WILDE, *Lady Windermere's Fan*

BIBLIOGRAPHY

PREFACE

Fimmel, R.O., W. Swindell, and E. Burgess. *Pioneer Odyssey.* NASA SP-396. Moffett Field, Calif.: Ames Research Center, 1977.

Perkins Dewey, A. *Robert Goddard: Space Pioneer.* Boston: Little, Brown, 1962.

Thom, A., and A.S. Thom. "A Megalithic Lunar Observatory in Orkney: The Ring of Brogar and Its Cairns." *Journal for the History of Astronomy,* Vol. IV, 1973, pp. 111–123.

CHAPTER 1

Beard, R.B., and A.C. Rotherham. *Space Flight and Satellite Vehicles.* London: Pitman, 1958.

Bergaust, E. *Wernher von Braun.* Washington, D.C.: National Space Institute, 1976.

Caidin, M. *Man into Space.* New York: Pyramid, 1961.

Clarke, A.C. *Interplanetary Flight.* New York: Harper & Bros., 1960.

______. *The Promise of Space.* New York: Harper & Row, 1968.

Cooper, H.S.F., Jr. *Apollo on the Moon.* New York: Dial, 1969.

Cox, D.W. *The Space Race: From Sputnik to Apollo and Beyond.* Philadelphia: Chilton, 1962.

Emme, E.M., ed. *Two Hundred Years of Flight in America: A Bicentennial Survey.* AAS History Series, Volume 1. San Diego: American Astronautical Society (Univelt, Inc.), 1977.

Flaherty, B.E., ed. *Psychophysiological Aspects of Space Flight.* New York: Columbia U. Press, 1961.

Gazenko, O.G., and H.A. Bjurstedt, ed. **Нукщиул и Лщсьщсу** (*Man in Space*). Moscow: **Гявфеукёсеищ "Рфнлф"** (Univelt, Inc., American distributor), 1974.

Hadingham, E. *Circles and Standing Stones.* New York: Walker, 1975.

Johnson, N.L. *Handbook of Soviet Manned Space Flight.* AAS Science and Technology Series, Vol. 48. San Diego: American Astronautical Society (Univelt, Inc.), 1980.

Ley, W. *Ranger to the Moon.* New York: Signet, 1965.

———. *Rockets, Missiles, and Men in Space.* New York, Viking, 1968.

———. *Rockets: The Future of Travel Beyond the Stratosphere.* New York: Viking, 1944.

Mailer, N. *Of a Fire on the Moon.* Boston: Little, Brown, 1970.

Mallan, L. *Men, Rockets and Space.* London: Cassell, 1956.

———. *Men, Rockets and Space Rats.* New York: Julian Messner, 1956.

Nock, O.S. *Railways Then and Now: A World History.* New York: Crown, 1975.

Rosen, M.L. *The Viking Rocket Story.* New York: Harper & Bros., 1956.

Setright, L.J.K. *The Designers: Great Automobiles and the Men Who Made Them.* Chicago: Follett, 1976.

Sharpe, M.R. *Living in Space.* New York: Doubleday, 1969.

Shepard, O. *The Lore of the Unicorn.* New York: Harper & Row, 1979.

Stein, R. *The Great Cars.* New York: Grosset & Dunlap, 1967.

Stine, H.G. *Earth Satellites.* New York: Ace, 1957.

Taylor, L.B., Jr. *Liftoff! The Story of America's Spaceport.* New York: Dutton, 1968.

Wolfe, T. *The Right Stuff.* New York: Bantam, 1979.

CHAPTER 2

Asimov, I. *Eyes on the Universe: A History of the Telescope.* Boston: Houghton Mifflin, 1975.

Ball, R.S. *Story of the Heavens.* London: Cassell, 1893.

Clark, R.W. *Einstein: The Life and Times.* New York: World, 1971.

Clerke, A.M. *A Popular History of Astronomy During the Nineteenth Century.* London: Adam & Charles Black, 1879.

Cole, F.W. *Fundamental Astronomy.* New York: Wiley, 1974.

Ionides, S.A., and M.L. Ionides. *Stars and Men.* New York: Bobbs-Merrill, 1939.

Jastrow, R., and M.H. Thompson. *Astronomy: Fundamentals and Frontiers.* New York: Wiley, 1977.

Van de Kamp, P. "The Nearby Stars." *Annual Review of Astronomy and Astrophysics,* Vol. 9, 1971, pp. 103–126.

______. "Unseen Astrometric Companions of Stars." *Annual Review of Astronomy and Astrophysics,* Vol. 13, 1975, pp. 295–333.

Vergon, V. *Relativity Beyond Einstein.* Los Angeles: Exeter Publications, 1976.

Wallace, A.R. *Man's Place in the Universe: A Study of the Results of Scientific Research in Relation to the Unity or Plurality of Worlds.* New York: McClure, Phillips, 1904.

Young, C.A. *A Textbook of General Astronomy.* Boston: Ginn & Co., 1898.

CHAPTER 3

Blaine, B.C.D. *The End of an Era in Space Exploration.* AAS Science and Technology Series, Vol. 42. San Diego: American Astronautical Society (Univelt, Inc.), 1976.

Brower, D. *The Starship and the Canoe.* New York: Holt, Rinehart, 1978.

Bussard, R.W., and R.D. DeLauer. *Nuclear Rocket Propulsion.* New York: McGraw-Hill, 1958.

Clark, J.D. *Ignition! An Informal History of Liquid Rocket Propellants.* New Brunswick, N.J.: Rutgers U. Press, 1972.

Cornelisse, J.W.; H.F.R. Schöyer, and K.F. Wakker. *Rocket Propulsion and Spaceflight Dynamics.* London: Pitman, 1979.

Dippery, D.F. "Matter-Antimatter Annihilation as an Energy Source in Propulsion." JPL Technical Memorandum 33-722. Pasadena, Calif.: Jet Propulsion Laboratory, 1977.

Dyson, F.J. "Death of a Project." *Science,* Vol. 149, 165, pp. 141–144.

Ferdman, S., ed. *The Second Fifteen Years in Space.* AAS Science and Technology Series, Vol. 31. San Diego: American Astronautical Society (Univelt, Inc.), 1973.

Fishback, J.F. "Relativistic Interstellar Spaceflight." *Astronautica Acta,* Vol. 15, 1969, pp. 25–35.

Heppenheimer, T.A. "On the Infeasibility of Interstellar Ramjets." *Journal of the British Interplanetary Society,* Vol. 31, 1978, pp. 54–55.

———. "Some Advanced Applications of a 1-million-second 1_{sp} Rocket Engine." *Journal of the British Interplanetary Society,* Vol. 28, 1975, pp. 173–188.

Hilton, J.I., and J.S. Luce. "Hypothetical Fusion Propulsion Rocket Vehicle." *Journal of Spacecraft,* Vol. 1, No. 3, 1964, pp. 276–282.

Holdren, J.P. "Fusion Energy in Context: Its Fitness for the Long Term." *Science,* Vol. 200, 1978, pp. 168–180.

Hsieh, T.M. "Thermonuclear Fusion Technology and Its Application in Space Propulsion." JPL Technical Memorandum 33-722. Pasadena, Calif.: Jet Propulsion Laboratory, 1977.

Johnston, R.S.; A. Naumann, Jr., and C.W. Fulcher. "The Future United States Space Program." *Advances in the Astronautical Sciences,* Vol. 38, Part I. San Diego: American Astronautical Society (Univelt, Inc.), 1979.

Kosmodeminsky, A.A. **Ярфьургепг Вукеуки Рфнлг Л.п. Ггщклщислгг.** Moscow: **Ищуррщу Гявфеукёсеищ Ьгргсеукзсеиф Щвщзрпш Сщюяф ССЗ**, 1954.

Lamore, L., and R.L. Gervais, eds. "Space Shuttles and Interplanetary Missions." *Advances in the Astronautical Sciences,* Vol. 28. San Diego: American Astronautical Society (Univelt, Inc.), 1970.

Ley, W. *Missiles, Moonprobes, and Megaparsecs.* New York: Signet, 1964.

———. *Rockets, Missiles, and Space Travel with Sputnik Data.* New York: Viking, 1957.

Martin, A.R. "The Effects of Drag on Relativistic Spaceflight." *Journal of the British Interplanetary Society,* Vol. 25, 1972, pp. 643–653.

———, ed. *Project Daedalus.* JBIS: Interstellar Studies. London: British Interplanetary Society, 1978.

Marx, G. "The Mechanical Efficiency of Interstellar Vehicles." *Astronautica Acta,* Vol. 9, 1963, pp. 131–139.

NASA. *Ion Propulsion for Spacecraft.* Cleveland: Lewis Research Center, 1977.

Nicolson, I. *The Road to the Stars.* New York: Morrow, 1978.

Nuckolls, J.; J. Emmett, and L. Wood. "Laser-Induced Thermonuclear Fusion." *Physics Today,* Vol. 26, 1973, pp. 46–53.

Nuckolls, J.; L. Wood, et al. "Laser Compression of Matter to Super-High Densities: Thermonuclear (CTR) Applications." *Nature,* Vol. 239, Sept. 1972, pp. 139–142.

Öpik, E.J. "Is Interstellar Travel Possible?" *Irish Astronomical Journal,* Vol. 6, No. 8, pp. 299–302.

Physics Today. "Space Vehicles Could Be Propelled by Remote Lasers," Vol. 30, Aug. 15, 1977, pp. 17, 19.

Papailiou, D.D. "Energy Exchange for Propulsion Application." JPL Technical Memorandum 33-722. Pasadena, Calif.: Jet Propulsion Laboratory, 1977.

______. "The Use of Matter-Antimatter Energy in Propulsion." JPL Technical Memorandum 33-722. Pasadena, Calif.: Jet Propulsion Laboratory, 1977.

Ragsdale, R.G. "To Mars in 30 Days by Gas-Core Nuclear Rocket." *Astronautics & Aeronautics,* Vol. 10, 1972, pp. 65–71.

Sänges, E. *Spaceflight: Countdown for the Future.* Ed. and tr. by K. Frucht. New York: McGraw-Hill, 1965.

Science. "Spacecraft Propulsion: New Methods." Vol. 176, 1972, pp. 167–178.

Science News. "Gravity's Repulsive Side." Vol. 117, No. 10, 1980, p. 48.

Sharpe, M., and F.I. Ordway. *The Rocket Team.* New York: Crowell, 1979.

Sloop, J. "Looking for the Sweet Combination." *Astronautics & Aeronautics,* Vol. 10, 1972, pp. 52–57.

Spencer, D.F., and L.D. Jaffe. "Feasibility of Interstellar Travel." *Astronautica Acta,* Vol. 9, 1963, pp. 49–58.

Steiner, D., and J.F. Clarke. "The Tokamak: Model T Fusion Reactor." *Science,* Vol. 199, 1978, pp. 1395–1403.

Stuhlinger, E. *Ion Propulsion for Space Flight.* New York: McGraw-Hill, 1964.

Thom, K. "Review of Fission Engine Concepts." *Journal of Spacecraft,* Vol. 9, No. 9, Sept. 1972, pp. 633–639.

Thomsen, D. "Reaching for the Stars." *Science News,* Vol. 116, 1979, pp. 52–54.

Ultraenergy Propellants: The Step Beyond High-energy Propulsion." *Chemical Week,* Oct. 12, 1957, pp. 148–156.

Vagners, J., ed. "The Outer Solar System." *Advances in the Astronautical Sciences,* Vol 29, Parts I and II. San Diego: American Astronautical Society (Univelt, Inc.), 1971.

CHAPTER 4

Asimov, I. *Extraterrestrial Civilizations.* New York: Crown, 1979.

Bartusiak, M.F. "Experimental Relativity." *Science News,* Vol. 116, Aug. 1979, pp. 140–142.

Bernstein, J. *Experiencing Science.* New York: Basic Books, 1978.

Berry, A. *The Iron Sun: Crossing the Universe Through Black Holes.* New York: Dutton, 1977.

Calder, N. *Einstein's Universe.* New York: Viking, 1979.

Conference on the Ancient Sun, Boulder, Col. *Fossil Record in the Earth, Moon and Meteorites.* Houston: Lunar and Planetary Institute, 1979.

Eardley, D.M. "Astrophysical Processes Near Black Holes." *Annual Review of Astronomy and Astrophysics,* Vol. 13, 1975, pp. 381–423.

Ford, K.W. "The World of Elementary Particles." *Research News.* Ann Arbor: U. of Michigan (newsletter).

Golden, F. *Quasars, Pulsars, and Black Holes: A Scientific Detective Story.* New York: Scribner's, 1976.

Gribbin, J. *White Holes: Cosmic Gushers in the Universe.* New York: Delacorte, 1977.

Jastrow, R. *Red Giants and White Dwarfs.* New York: Harper & Row, 1976.

Science News.

"Axion Hunt: Getting Something Out of Nothing." Vol. 113, No. 15, April 15, 1979, p. 229.

"Bare Bottom, Naked Charm: Booms-a-daisy Physics." Vol. 116, Sept. 22, 1979, pp. 196–197.

"Physics Is an Abstract Art." Vol. 113, No. 5, Feb. 4, 1978, p. 68.

"Summer of the Gluons: Jets of Evidence." Vol. 116, Sept. 1, 1979, pp. 151–152.

"Weighed in the Balance and Found: Neutrino." Vol. 117, May 10, 1980, pp. 292–293.

Shklovskii, I.S. *Stars: Their Birth, Life, and Death.* Tr. by Richard B. Rodman. San Francisco: W.H. Freeman, 1978.

Slater, N.B. *The Development and Meaning of Eddington's "Fundamental Theory."* Cambridge, England: The University Press, 1957.

Thomsen, D.E. "An Angle on the New Physics." *Science News,* Vol. 116, Dec. 15, 1979, pp. 409, 412.

______. "Leapin' Leptons." *Science News,* Vol. 115, Jan. 20, 1979, pp. 42–43.

______. "Physics with Strings Attached." *Science News,* Vol. 116, Oct. 20, 1979, pp. 266–267.

Trefil, J.S. "It's all relative when you travel faster than light." *Smithsonian,* Nov. 1978, pp. 132–138.

Wald, R.M. *Space, Time, and Gravity: The Theory of the Big Bang and Black Holes.* Chicago: U. of Chicago Press, 1977.

Wilford, J.N., ed. *Scientists at Work: The Creative Process of Scientific Research.* New York: Dodd, Mead, 1979.

CHAPTER 5

Cohen, S.N. "Recombinant DNA: Fact and Fiction." *Science,* Vol. 196, 1977, pp. 407–408.

Crowe, J.H., and A.F. Cooper, Jr. "Cryptobiosis." *Scientific American,* 1975, pp. 30–36.

Flaherty, B.E., ed. *Psychophysiological Aspects of Space Flight.* New York: Columbia U. Press, 1961.

Fuller, J.G. *Fever: The Hunt for a New Killer Virus.* New York: Reader's Digest, 1974.

Hayflick, L. "The Cell Biology of Human Aging." *Scientific American,* Vol. 242, No. 1, pp. 42–49.

Heinlein, R.A. *Orphans of the Sky.* New York: Berkley, 1963.

Miller, J.A. "Puzzling Out the Cell's Power Plant." *Science News,* Vol. 116, Sept. 15, 1979, pp. 184–185.

Schmeck, H.M., Jr. "Cloned Antibodies Promise Medical Revolution." *The New York Times*, Aug. 5, 1980, p. C1.

Science News.

"Cell's Gene-Copying Machinery Anchored." Vol. 117, Mar. 1, 1980, p. 132.

"The Cloning of a Man: Debate Begins." Vol. 113, Mar. 18, 1978, p. 164.

"DNA on X Chromosome Cloned and Identified." Vol. 118, Aug. 30, 1980, p. 132.

"Double Helix Reveals New Twist." Vol. 116, Dec. 22 & 29, 1979, pp. 420–421.

"Gene Rules: Violations and Revisions." Vol. 112, Dec. 24 & 31, 1977, p. 420.

"Mouse-to-Mouse Gene Transfer." Vol. 117, April 19, 1980, p. 244.

"On the Way to a Clone." Vol. 116, July 28, 1979, p. 68.

"Two Mothers, No Father = One Embryo." Vol. 116, Aug. 18, 1979, pp. 116–117.

Sharpe, M.R. *Living in Space: The Astronaut and His Environment.* New York: Doubleday, 1969.

Tipler, F.J. "Extraterrestrial Intelligent Beings Do Not Exist." *Quarterly Journal of the Royal Astronomical Society*, Vol. 21, 1980, pp. 267–281.

Watson, J.D. *The Double Helix: A Personal Account of the Discovery of the Structure of DNA.* New York: Atheneum, 1968.

CHAPTER 6

American Astronautical Society. *A Speculative Analysis Correlating Stellar Spectral Classes with Psychological*

Base of Potential Civilizations. AAS 69-387-2 (Part 2). San Diego: American Astronautical Society, 1969.

Anders, E.; R. Hayatsu, and M.H. Studier. "Organic Compounds in Meteorites." *Science,* Vol. 182, Nov. 23, 1973, pp. 781–789.

Burns, J.A., ed. *Planetary Satellites.* Tucson: U. of Arizona Press, 1977.

Christian, J.L., ed. *Extra-Terrestrial Intelligence: The First Encounter.* Buffalo: Prometheus Books, 1976.

Clement, H. *Mission of Gravity.* New York: Doubleday, 1954.

Cole, F.W. *Fundamental Astronomy: Solar System and Beyond.* New York: Wiley, 1974.

Cole, G.H.A. *The Structure of Planets.* New York: Crane, Russak, 1978.

Cook, A.H. *Physics of the Earth and Planets.* New York: Wiley, 1973.

Derham, W. *Astro-Theology or a Demonstration of the Being and Attributes of God, from a Survey of the Heavens.* London: W. & J. Innys, 1719.

Dole, S.H. *Habitable Planets for Man.* New York: Blaisdell Publishing, 1964.

———, and I. Asimov. *Planets for Man.* New York: Random House, 1964.

Eberhart, J. "Giving Ourselves Away." *Science News,* Vol. 113, No. 9, Mar. 4, 1978, pp. 138–139.

Foster, G.V. "Non-Human Artifacts in the Solar System." *Spaceflight,* Vol. 14, 1972, pp. 447–453.

Garmon, L. "The Empyrean Strikes Back." *Science News,* June 14, 1980, p. 381.

Gehrels, T., ed. *Protostars and Planets: Studies of Star Formation and of the Origin of the Solar System.* Tucson: U. of Arizona Press, 1978.

Hahn, H.M. *Erde, Sonne und Planeten: Raumsonden Erforschen des Sonnensystem.* Cologne: Kiepenheuer & Witsch, 1978.

Hart, M.H. "The Evolution of the Atmosphere of the Earth." *Icarus*, Vol. 33, 1978, pp. 23–39.

Hartmann, W.K. "Planet Formation: Mechanism of Early Growth." *Icarus*, Vol. 33, 1978, pp. 50–61.

Heppenheimer, T.A. *Toward Distant Suns.* Harrisburg, Pa.: Stackpole Books, 1979.

McAlister, H.A. "Speckle Interferometry as a Method for Detecting Nearby Extrasolar Planets." *Icarus*, Vol. 30, 1977, pp. 789–792.

Maunder, E.W. *Are the Planets Inhabited?* London: Harper & Bros., 1913.

Molton, P.M. "On the Likelihood of a Human Interstellar Civilization." *Journal of the British Interplanetary Society*, Vol. 31, 1978, pp. 203–208.

Murray, B.; S. Gulkis, and R.E. Edelson. "Extraterrestrial Intelligence: An Observational Approach." *Science*, Vol. 199, No. 4328, 1978, pp. 485–491.

Narrien, J. *Historical Account of the Origin and Progress of Astronomy.* London: Baldwin & Cradock, 1833.

Powell, C. "Interstellar Flight and Intelligence in the Universe." *Spaceflight*, Vol. 14, 1972, pp. 442–447.

Proctor, R.A. *Other Worlds Than Ours: The Plurality of Worlds Studies Under the Light of Recent Scientific Research.* New York: D. Appleton, 1897.

Sagan, C., ed. *Communication with Extraterrestrial In-*

telligence: CETI. Cambridge: MIT Press, 1973.

Science. "A Search for Ultra-Narrowband Signals of Extraterrestrial Origin." Vol. 201, August 1978, pp. 733–735.

Science News. "Mass Extinction: More Theories." Vol. 116, Nov. 24, 1979, p. 356.

Tombaugh, C.W., and P. Moore. *Out of the Darkness: The Planet Pluto.* Harrisburg, Pa.: Stackpole Books, 1980.

Van de Kamp, P. "The Nearby Stars." *Annual Review of Astronomy and Astrophysics,* Vol. 9, 1971, pp. 103–126.

______. "Unseen Astrometric Companions of Stars." *Annual Review of Astronomy and Astrophysics,* Vol. 13, 1975, pp. 295–333.

Whewell, W. *The Plurality of Worlds.* Boston: Gould & Lincoln, 1854.

CHAPTER 7

Baker Fahnestock, W., M.D. *Worlds Within Worlds, Discoveries in Astronomy: The Sun and Stars Inhabited.* Philadelphia: Barclay, 1876.

Bond, A. "Problems of Interstellar Propulsion." *Spaceflight,* Vol. 13, 1971, pp. 245–251.

Despain, L.G.; J.P. Hennes, and J.L. Archer. "Scientific Goals of Missions Beyond the Solar System." AAS 71-163. *Advances in the Astronautical Sciences,* Vol. 29, Part II, pp. 597–616. San Diego: American Astronautical Society (Univelt, Inc.), 1971.

Dessler, A.J., and R.A. Park. "The First Step Beyond the

Solar System." AAS 71-165. *Advances in the Astronautical Sciences,* Vol. 29, Part II, pp. 681–690. San Diego: American Astronautical Society (Univelt, Inc.), 1971.

Ehricke, K.A. "The Ultraplanetary Probe." AAS 71-164. *Advances in the Astronautical Sciences,* Vol. 29, Part II, pp. 617–680. San Diego: American Astronautical Society (Univelt, Inc.), 1971.

Gatewood, G. "On the Astrometric Detection of Neighboring Planetary Systems." *Icarus,* Vol. 27, 1976, pp. 1–12.

Hart, M.H. "Habitable Zones About Main Sequence Stars." *Icarus*, Vol. 37, 1979, pp. 351–357.

Heppenheimer, T.A. "On the Formation of Planets in Binary Star Systems." *Astronomy and Astrophysics,* Vol. 65, 1978, pp. 421–426.

———. "Some Advanced Applications of a 1-Million-Second 1_{sp} Rocket Engine." *Journal of the British Interplanetary Society,* Vol. 28, 1975, pp. 175–181.

Hunter, M.H., II. "Accessible Regions Beyond the Solar System." AAS69-386. *Advances in the Astronautical Sciences,* Vol. 26, pp. 293–317. San Diego: American Astronautical Society (Univelt, Inc.), 1970.

Martin, A.R., ed. *Project Daedalus.* London: British Interplanetary Society, 1978.

Martin, A.R., "The Detection of Extrasolar Planetary Systems":
Part I: "Methods of Detection." *Journal of the British Interplanetary Society,* Vol. 27, 1974, pp. 643–659.

Part II: "Discussion of Astrometric Results." *JBIS,* Vol. 27, 1974, pp. 881–906.

Part III: "Review of Recent Developments." *JBIS,* Vol. 28, 1975, pp. 182–190.

NASA. *Project Orion: A Design Study of a System for Detecting Extrasolar Planets.* NASA SP-436. Moffett Field, Calif.: Ames Research Center, 1980.

_____. *The Voyage of Mariner 10: Mission to Venus and Mercury.* NASA SP-424. Pasadena, Calif.: Jet Propulsion Laboratory, 1978.

Powers, R.M. *Planetary Encounters: The Future of Unmanned Spaceflight.* Harrisburg, Pa.: Stackpole Books, 1978.

Science. "Gravitational Lens of the Sun: Its Potential for Observations and Communication Over Interstellar Distances." Vol. 205, 1979, pp. 1133–1135.

Thomsen, D.E. "Multiple Mirror Astronomy." *Science News,* Vol. 118, Aug. 16, 1980, pp. 106–108.

_____. "Will Astronomy Go Into Orbit?" *Science News,* Vol. 118, Aug. 30, 1980, pp. 138–140.

Von Braun, W. *The Mars Project.* Urbana: U. of Illinois Press, 1962.

CHAPTER 8

Agosto, W.N. "Accessible Industrial Materials in Lunar Soil." AAS 78-190. *Advances in the Astronautical Sciences,* Vol. 38, Part II, pp. 369–384. San Diego: American Astronautical Society (Univelt, Inc.), 1979.

Bauer, H.E., and D.E. Lockwood. "The Case for an Early Manned Lunar Orbit Station." AAS 70-022. *Advances in the Astronautical Sciences,* Vol. 27, pp. 111–132. San Diego: American Astronautical Society (Univelt, Inc.), 1970.

Bekey, I., and H. Mayer. "1980–2000: Raising Our Sights for Advanced Space Systems." *Astronautics & Aeronautics,* Vol. 14, 1976, pp. 34–63.

Cole, D.M., and D.W. Cox. *Islands in Space.* Philadelphia: Chilton, 1964.

Eberhart, J. "Sunsat: Collecting Solar Power in Orbit." *Science News,* Vol. 113, No. 16, 1979, pp. 256–257.

Ehricke, K.A. "Earth-Moon Transportation." AAS 70-058. *Advances in the Astronautical Sciences,* Vol. 27, pp. 401–452. San Diego: American Astronautical Society (Univelt, Inc.), 1970.

Helin, E.F., and E.M. Shoemaker. "Discovery of Asteroid AA[1]." *Icarus,* Vol. 31, 1977, pp. 415–419.

Heppenheimer, T.A. *Colonies in Space.* Harrisburg, Pa.: Stackpole Books, 1977.

Herman, D. "Planetary Exploration Objectives 1980–1990." AAS 78-185. *Advances in the Astronautical Sciences,* Vol. 38, Part I, p. 361. San Diego: American Astronautical Society (Univelt, Inc.), 1979.

Kingsbury, D. "A Hybrid Chemical Nuclear Space Freighter Concept." AAS 77-219. *Advances in the Astronautical Sciences,* Vol. 36, Part II, pp. 139–156. San Diego: American Astronautical Society (Univelt, Inc.), 1978.

Kleiman, L.A., ed. *Project Icarus.* Cambridge, Mass.: MIT Press, 1967.

Kline, R. "Space Structure: A Key to New Opportunities." AAS 79-059. Science and Technology Series, Vol. 49, 1980, pp. 161–174.

Knight, D.C. *The Tiny Planets.* New York: Morrow, 1973.

Molton, P.M., and T.E. Divine. "The Use of Outer Planet Satellites and Asteroids as Sources of Raw Materials for Life Support Systems." AAS 77-236. *Advances in the Astronautical Sciences,* Vol. 36, Part I, pp. 309–348. San Diego: American Astronautical Society (Univelt, Inc.), 1978.

Niehoff, J. "An Assessment of Comet and Asteroid Missions." AAS 71-104. *Advances in the Astronautical Sciences,* Vol. 29, Part I, pp. 93–120. San Diego: AAS, 1971.

O'Neill, G.K., and H.H. Kolm. "Mass Driver for Lunar Transport and as Reactor Engine." AAS 77-222. *Advances in the Astronautical Sciences,* Vol. 36, Part 1, pp. 191–208. San Diego: AAS, 1978.

Pilcher, F., and J. Meeus. *Tables of Minor Planets.* Cincinnati: U. of Cincinnati, 1973.

Powers, R.M. *Shuttle: The World's First Spaceship.* Harrisburg, Pa.: Stackpole Books, 1979.

Riedesel, R.G., and R.S. Cowls. "Manned Planetary Missions with Reusable Nuclear Shuttles." AAS 70-040. *Advances in the Astronautical Sciences,* Vol. 28, pp. 457–473. San Diego: AAS, 1970.

Robey, D.H. "A Theory on the Nature and Origin of Comets with Implications for Space Mission Planning." AAS 70-029. *Advances in the Astronautical Sciences,* Vol. 28, pp. 259–302. San Diego: AAS, 1970.

Roth, G.D. *The System of Minor Planets.* Princeton, N.J. Van Nostrand, 1962.

Russak, S.L. "Design of an Advanced Meteoroid Probe of the Asteroid Belt." AAS 69-321. *Advances in the Astronautical Sciences,* Vol. 26, p. 289. San Diego: AAS, 1970.

Wetherill, G.W. "Apollo Objects." *Scientific American,* March 1979, pp. 54–65.

Wrobel, J.R., and D.J. Kerrisk. "Early Exploration of the Asteroids Region by Solar Powered Electrically Propelled Spacecraft." AAS 69-322. *Advances in the Astronautical Sciences,* Vol. 26, pp. 267–289. San Diego: AAS, 1970.

CHAPTER 9

Agosto, W.N. "Industrial Materials in Lunar Soil." AAS 78-190. *Advances in the Astronautical Sciences,* Vol. 38, Part I, pp. 369–383. San Diego: AAS, 1979.

Anderson, P. "Scenarios of the Future in Space in the Year 2069." AAS 79-313. Science and Technology Series, Vol. 50, 1980, pp. 69–81.

Bauer, H.E., and D.E. Lockwood. "The Case for an Early Manned Lunar Orbit Station." AAS 70-022. *Advances in the Astronautical Sciences,* Vol. 27, pp. 111–131. San Diego: AAS, 1970.

Billman, K. "Solares Orbiting Mirror System." AAS 79-304. Science and Technology Series, Vol. 50, 1980, pp. 15–25.

Bluth, B.J. "Constructing Space Communities: A Critical Look at the Paradigms." AAS 78-141. *Advances in the Astronautical Sciences,* Vol. 38, Part II, pp. 453–475. San Diego: AAS, 1979.

Bock, E.H.; F. Lambrou, and M. Simon, Jr. "Habitat Design—an Update." AAS 77-272. *Advances in the Astronautical Sciences,* Vol. 36, Part I, pp. 575–590. San Diego: AAS, 1978.

Calder, N. *Spaceships of the Mind.* New York: Viking, 1978.

Cole, D.M., and D.W. Cox. *Islands in Space.* Philadelphia: Chilton, 1964.

Criswell, D.R., and R.D. Waldon. "Utilization of Lunar Materials in Space." AAS 78-198. *Advances in the Astronautical Sciences,* Vol. 38, Part II, p. 841. San Diego: AAS, 1978.

Engler, E.E., and W.K. Muench. "Automated Space Fabrication of Structural Elements." AAS 77-200. *Ad-*

vances in the Astronautical Sciences, Vol. 36, Part 1, pp. 27–55. San Diego: AAS, 1978.

Green, J. "The Polar Lunar Base—A Viable Alternative to L-5." AAS 78-191. *Advances in the Astronautical Sciences,* Vol. 38, Part I, pp. 385–425. San Diego: AAS, 1979.

Halbouty, M.T. "Future Programs and Prospects for Resource Exploration from Space by the Year 2000." AAS 78-182. *Advances in the Astronautical Sciences,* Vol. 38, Part II, pp. 721–239. San Diego: AAS, 1979.

Heppenheimer, T.A. *Colonies in Space.* Harrisburg, Pa.: Stackpole Books, 1977.

______. *Toward Distant Suns.* Harrisburg, Pa.: Stackpole Books, 1979.

Holmen, R.E., and F.C. Runge. "Operational Concepts for a 10-Year Space Station." AAS 70-031. *Advances in the Astronautical Sciences,* Vol. 27, pp. 173–223. San Diego: AAS, 1970.

Hudson, G.C. "Advanced Propulsion Systems and Solar System Spaceships." AAS 79-308. Science and Technology Series, Vol. 50, 1980, pp. 47–60.

Kober, C.L. "Commercial Use of Space Station." AAS 70-036. *Advances in the Astronautical Sciences,* Vol. 27, pp. 299–301. San Diego: AAS, 1970.

Kornberg, J.P.; P.K. Chapman, and P.E. Glaser. "Health Maintenance and Health Surveillance Considerations for an SPS Space Construction Base Community." AAS 78-176. *Advances in the Astronautical Sciences,* Vol. 38, Part II, pp. 651–661. San Diego: AAS, 1979.

Ley, W. *Rockets, Missiles, and Space Travel.* New York: Viking, 1957.

NASA. *Space Resources and Space Settlements.* NASA SP-428. Moffett Field, Calif.: Ames Research Center, 1979.

______. *Space Settlements: A Design Study.* NASA SP-413. Moffett Field, Calif.: Ames Research Center, 1977.

Naumann, R.J. "Materials Processing in Space: A Strategy for Commercialization." AAS 77-238. *Advances in the Astronautical Sciences,* Vol. 36, Part I, pp. 349–364. San Diego: AAS, 1978.

O'Neill, G.K. *The High Frontier: Human Colonies in Space.* New York: Morrow, 1977.

Powell, D.J., and L. Browning. "Automated Fabrication of Large Space Structures." *Aeronautics & Astronautics,* October 1978, pp. 24–59.

Ryan, C., ed. *Across the Space Frontier.* New York: Viking, 1952.

Shapira, J. "Recent Advances in Closed Support System Concepts." AAS 69-143. *Advances in the Astronautical Sciences,* Vol. 26, pp. 107–117. San Diego: AAS, 1970.

Smith, R.E., D.L. Erickson, and W.P. Pratt. "Ten-Year Space Station Experiment Program." AAS 70-033. *Advances in the Astronautical Sciences,* Vol. 27, pp. 225-255. San Diego: AAS, 1970.

Spurlock, J.M., and M. Modell. "Technology Requirements for Closed-Ecology Life Support Systems Applicable to Space Habitats." AAS 77-273. *Advances in the Astronautical Sciences,* Vol. 36, Part II, pp. 527–545. San Diego: AAS, 1978.

Toerge, F., and C.A. O'Donnell. "Space Station Habitability: Its Form Relationship to Man." AAS 70-023. *Advances in the Astronautical Sciences,* Vol. 27, pp. 163–171. San Diego: AAS, 1970.

Trotti, G., and D.R. Criswell. "Lunar Base Habitat Exploration Design Study." AAS 78-143. *Advances in the Astronautical Sciences,* Vol. 38, Part II, p. 477. San Diego: AAS, 1979.

Vajk, J.P. "Planetary Exploration Space Colony Style." AAS 79-307. Science and Technology Series, Vol. 50, pp. 35–46. San Diego: AAS, 1980.

CHAPTER 10

Bock, E.H.; F. Lambrou, Jr., and M. Simon. "Habitat Design—An Update." AAS 77-272. *Advances in the Astronautical Sciences,* Vol. 36, Part II, pp. 575–590. San Diego: AAS, 1978.

Clarke, A.C. *Rendezvous with Rama.* New York: Harcourt, Brace, 1973.

Cole, D.M., and D.W. Cox. *Islands in Space.* Philadelphia: Chilton, 1964.

Hamilton, W.T. "Engineering in the 21st Century." AAS 78-194. *Advances in the Astronautical Sciences,* Vol. 38, Part II, pp. 783–791. San Diego: AAS, 1979.

Hawley, P.A.M. "Optical and Electronic Imaging Systems for Solar System Exploration." AAS 71-115. *Advances in the Astronautical Sciences,* Vol. 29, Part I, pp. 331–339. San Diego: AAS, 1971.

Heppenheimer, T.A. *Colonies in Space.* Harrisburg, Pa.: Stackpole Books, 1977.

Kase, P.G. "Radiation Model of Man for Analysis of Future Space Missions." AAS 70-054. *Advances in the Astronautical Sciences,* Vol. 27, pp. 513–549. San Diego: AAS, 1970.

McAlister, H.A. "Speckle Interferometry as a Method for

Detecting Nearby Extrasolar Planets." *Icarus,* Vol. 30, 1977, pp. 789–792.

McCarthy, J.F., Jr. "Engineering in the 21st Century." AAS 78-192. *Advances in the Astronautical Sciences,* Vol. 38, Part I, pp. 757–771. San Diego: AAS, 1979.

MacManus, S. *The Story of the Irish Race.* Old Greenwich, Ct.: Devin-Adair, 1979.

NASA. *Space Resources and Space Settlements.* NASA SP-428. Moffett Field, Calif.: Ames Research Center, 1979.

______. *Space Settlements: A Design Study.* NASA SP-413. Moffett Field, Calif.: Ames Research Center, 1977.

O'Neill, G.K. *The High Frontier.* New York: Morrow, 1977.

Paluszek, M.A. "Magnetic Radiation Shielding for Permanent Space Habitats." AAS 77-274. *Advances in the Astronautical Sciences,* Vol. 36, Part I, pp. 545–575. San Diego: AAS, 1978.

Phillips, J.M.; A.D. Harlan, and K.C. Krumhar. "Developing Closed Life Support Systems for Large Space Habitats." AAS 78-145. *Advances in the Astronautical Sciences,* Vol. 38, Part II, pp. 517–547. San Diego: AAS, 1979.

Shapira, J. "Recent Advances in Closed Life Support System Concepts." AAS 69-143. *Advances in the Astronautical Sciences,* Vol. 26, pp. 107–118. San Diego: AAS, 1970.

Sheckley, R. *Futuropolis.* New York: A & W Visual Library, 1978.

Spurlock, J.M., and M. Modell. "Systems Applicable to Space Habitats." AAS 77-273. *Advances in the Astro-*

nautical Sciences, Vol. 36, Part I, pp. 527–545. San Diego: AAS, 1978.

Trotti, G., and D.R. Criswell. "Lunar Base Habitat Exploration Design Study." AAS 78-143. *Advances in the Astronautical Sciences,* Vol. 38, Part II, p. 477. San Diego: AAS, 1979.

Ziegler, W. "A Preliminary Investigation of Space Habitat Atmospheres." AAS 77-284. *Advances in the Astronautical Sciences,* Vol. 36, Part II, pp. 961–982. San Diego: AAS, 1978.

CHAPTER 11

Akins, F.R. "Isolation and Confinement: Considerations for Colonization." AAS 77-245. *Advances in the Astronautical Sciences,* Vol. 36, Part II, pp. 731-750. San Diego: AAS, 1978.

Bluth, B.J. "Astronaut Stress—Shuttle/Space Work Environment." AAS 79-316. Science and Technology Series, Vol. 50, 1980, pp. 95–110.

Cooper, H.S.F., Jr. *A House in Space.* New York: Holt, 1976.

Fregger, B. "Earthward Implications of Cosmic Migration." AAS 79-322. Science and Technology Series, Vol. 50, 1980, pp. 167–172.

Gorove, S. "Legal Ties of a Space Colony to Earth." AAS 77-260. *Advances in the Astronautical Sciences,* Vol. 36, Part II, pp. 803–808. San Diego: AAS, 1978.

Mack. R., and T. Cullinan. "Space Community Planning in a Down-to-Earth Context." AAS 77-280. *Advances in the Astronautical Sciences,* Vol. 36, Part II, pp. 943–948. San Diego: AAS, 1978.

McWilliams, R.D. "On the Possible Sociopolitical Nature of the Early L-5 Space Colonies: An Ecological Approach." AAS 78-140. *Advances in the Astronautical Sciences,* Vol. 38, Part II, pp. 429–452. San Diego: AAS, 1979.

Maruyama, M. "Design Principles and Cultures." AAS 77-282. *Advances in the Astronautical Sciences,* Vol. 36, Part II, pp. 949–960. San Diego: AAS, 1978.

Sells, S.B., and E.K. Gunderson. "A Social System Approach to Long-Duration Missions." *Human Factors in Long-Duration Spaceflight.* Washington, D.C.: National Academy of Sciences, 1972.

CHAPTER 12

Aviation Week & Space Technology. "Self-Healing Computer in Development." Aug. 15, 1977, pp. 57–60.

Copeland, B. "Communication Requirements of a Space Settler: The Need to Keep in Touch." AAS 77-246. *Advances in the Astronautical Sciences,* Vol. 36, Part II, pp. 751–764. San Diego: AAS, 1978.

Heppenheimer, T.A. "Some Advanced Applications of a 1-Million-Second 1_{sp} Rocket Engine." *Journal of the British Interplanetary Society,* Vol. 28, 1975, pp. 175–181.

Hoag, D.G., and W. Wrigley. "Navigation and Guidance in Interstellar Space." AAS 79-328. Science and Technology Series, Vol. 50, 1980, pp. 197–202.

Johnson, M.L. "Long Duration Mission Reliability via Multiprocessing." AAS 71-158. *Advances in the Astronautical Sciences,* Vol. 29, Part II, pp. 495–509. San Diego: AAS, 1971.

Martin, A.R. "The Effects of Drag on Relativistic Spaceflight." *Journal of the British Interplanetary Society,* Vol. 25, 1972, pp. 643–653.

Martin, A.R., ed. *Project Daedalus.* London: British Interplanetary Society, 1978.

Matloff, G.L. "Utilization of O'Neill's Model I Lagrange Point Colony as an Interstellar Ark." *Journal of the British Interplanetary Society,* Vol. 29, 1976, pp. 775–785.

Noyce, R.N. "Microelectronics." *Scientific American,* Vol. 237, No. 3, Sept. 1977, pp. 63–69.

Sagan, C., ed. *Communication with Extraterrestrial Intelligence: CETI.* Cambridge, Mass.: MIT Press, 1976.

Stumpf, C.L. "Reliability Improvement Techniques for Long Life Missions." AAS 71-156. *Advances in the Astronautical Sciences,* Vol. 29, Part II, pp. 451–468, 1971.

Thomsen, D.E. "Convocation to Contemplate Quasars." *Science News,* Vol. 116, Sept. 29, 1979, pp. 217–221.

White, C.F., and P.M. Hooten. "Design Considerations for Improvement in Spacecraft Reliability for Long Duration Space Missions." AAS 71-155. *Advances in the Astronautical Sciences,* Vol. 29, Part II, pp. 437–450. San Diego: AAS, 1971.

White, R.M. "Disk-Storage Technology." *Scientific American,* Aug. 1980, pp. 138–148.

CHAPTER 13

Archer, J.L., and A.J. O'Donnell. "The Scientific Exploration of Near Stellar Systems." *Advances in the Astronautical Sciences,* Vol. 29, Part II, pp. 691–724. San Diego: AAS, 1971.

Dole, S.H., and I. Asimov. *Planets for Man.* New York: Random House, 1964.

Edelson, E. *Who Goes There? The Search for Intelligent Life in the Universe.* New York: Doubleday, 1979.

Heppenheimer, T.A. *Toward Distant Suns.* Harrisburg, Pa.: Stackpole Books, 1979.

Masursky, H. "Radar Imagery and Future Exploration of Venus." AAS 78-186. *Advances in the Astronautical Sciences,* Vol. 38, Part II, p. 362. San Diego: AAS, 1979.

Molton, P.M. "On the Likelihood of a Human Interstellar Civilization." *Journal of the British Interplanetary Society,* Vol. 31, 1978, pp. 203–308.

Powers, R.M. *Planetary Encounters.* New York: Warner, 1980.

Richards, G.R. "Planetary Detection from an Interstellar Probe." *Journal of the British Interplanetary Society,* Vol. 21, 1975, pp. 579–585.

Science. "Gravitational Lens of the Sun: Its Potential for Observations and Communications Over Interstellar Distances." Vol. 205, 1979, pp. 1133–1135.

Trefil, J.S. "It's all relative when you travel faster than light." *Smithsonian,* Nov. 1978, pp. 132–139.

CHAPTER 14

Jones, E.M. "Colonization of the Galaxy." *Icarus,* Vol. 28, 1978, pp. 245–251.

——. "Interstellar Colonization." *Journal of the British Interplanetary Society,* Vol. 31, pp. 103–107.

Oberg, J. "Terraforming." *Astronomy,* May 1978, pp. 7–25.

INDEX